T0185615

This series fosters information exchange and discussion on all aspects of manufacturing and surface engineering for modern industry. This series focuses on manufacturing with emphasis in machining and forming technologies, including traditional machining (turning, milling, drilling, etc.), non-traditional machining (EDM, USM, LAM, etc.), abrasive machining, hard part machining, high speed machining, high efficiency machining, micromachining, internet-based machining, metal casting, joining, powder metallurgy, extrusion, forging, rolling, drawing, sheet metal forming, microforming, hydroforming, thermoforming, incremental forming, plastics/composites processing, ceramic processing, hybrid processes (thermal, plasma, chemical and electrical energy assisted methods), etc. The manufacturability of all materials will be considered, including metals, polymers, ceramics, composites, biomaterials, nanomaterials, etc. The series covers the full range of surface engineering aspects such as surface metrology, surface integrity, contact mechanics, friction and wear, lubrication and lubricants, coatings an surface treatments, multiscale tribology including biomedical systems and manufacturing processes. Moreover, the series covers the computational methods and optimization techniques applied in manufacturing and surface engineering. Contributions to this book series are welcome on all subjects of manufacturing and surface engineering. Especially welcome are books that pioneer new research directions, raise new questions and new possibilities, or examine old problems from a new angle. To submit a proposal or request further information, please contact Dr. Mayra Castro, Publishing Editor Applied Sciences, via mayra.castro@springer.com or Professor J. Paulo Davim, Book Series Editor, via pdavim@ua.pt

More information about this subseries at http://www.springer.com/series/10623

Fredrick Madaraka Mwema ·
Esther Titilayo Akinlabi

Fused Deposition Modeling

Strategies for Quality Enhancement

 Springer

Fredrick Madaraka Mwema🆔
Department of Mechanical Engineering
Dedan Kimathi University of Technology
Nyeri, Kenya

Esther Titilayo Akinlabi🆔
Department of Mechanical
Engineering Science
University of Johannesburg
Auckland Park, Johannesburg, South Africa

ISSN 2191-530X   ISSN 2191-5318 (electronic)
SpringerBriefs in Applied Sciences and Technology
ISSN 2365-8223   ISSN 2365-8231 (electronic)
Manufacturing and Surface Engineering
ISBN 978-3-030-48258-9   ISBN 978-3-030-48259-6 (eBook)
https://doi.org/10.1007/978-3-030-48259-6

This Springer imprint is published by the registered company Springer Nature Switzerland AG
The registered company address is: Gewerbestrasse 11, 6330 Cham, Switzerland

Preface

Fused deposition modelling (FDM) is one of the most progressive and advanced Additive Manufacturing (AM) methods for the modern industry. The method has proven its suitability as a rapid prototyping technique and for the production of intricate functional components. However, the process faces two major limitations; poor surface quality and limited range of materials as it is mostly applicable to polymer-based raw materials. There are therefore extensive efforts by the AM researchers to enhance the quality of FDM parts and expand its application in different fields.

This book contributes to these continued efforts by presenting different strategies for quality enhancement of the technology. In the book, the terms FDM and 3D printing have been used interchangeably and they have the same meaning. The book is presented in four chapters. In Chap. 1, a general introduction to the fused deposition modelling is presented. A glimpse into different methods of AM technology, science of FDM and its applications, process parameters and quality aspects of FDM technology are presented in this chapter. Most importantly, the role of 3D printing in the fight against Coronavirus disease of 2019 is discussed in Chap. 1. In Chap. 2, a full factorial approach for the design of experiment (DOE) based on different levels of print orientation and layer resolution during FDM of PLA simple samples is presented as a strategy for enhancing both surface finish and micro-hardness properties. In Chap. 3, a multi-objective optimization approach is presented as another strategy for quality enhancement of FDM parts using case studies both from the literature and the experimental work by the authors. Finally, surface engineering technology is presented as a strategy for enhancing the surface and functional quality of the FDM parts in Chap. 4.

All the borrowed information such as methods, data and figures have been acknowledged accordingly inside the text. It is the hope of the authors that the book will, holistically, contribute towards expanding applications of fused deposition modelling parts. The book is suitable for engineers, researchers, academics and industrialists in the 3D printing field.

We acknowledge Springer for accepting to publish this work and the professional support they have accorded to us in the course of developing the idea and writing of the manuscript.

Nairobi, Kenya   Fredrick Madaraka Mwema
Johannesburg, South Africa   Esther Titilayo Akinlabi

Contents

About the Authors

Fredrick Madaraka Mwema Dr. F. M. Mwema is a postdoctoral researcher and a lecturer at the University of Johannesburg, South Africa and Dedan Kimathi University of Technology, Kenya, respectively. He is currently the Head of Department of Mechanical Engineering Department at Dedan Kimathi University, Kenya. He obtained BSc and MSc degrees in Mechanical Engineering from Jomo Kenyatta University of Agriculture and Technology (JKUAT), Kenya, in 2011 and 2015, respectively. He has a PhD in Mechanical Engineering from the University of Johannesburg, which he obtained in 2019. His PhD research work involved thin film coatings for surface protection and functional components. He has interests in advanced manufacturing, severe plastic deformation processes, additive manufacturing, thin film depositions, surface engineering, and materials characterizations. In thin films, Dr. Mwema has interest in fractal theory of coatings for enhanced depositions and behaviour in advanced applications. He has published more than 50 articles in peer-reviewed journals, conferences, and book chapters. He supervises and mentors several students, currently with 4 masters and 3 PhD students. He has over 6 years of university teaching and training experience in mechanical engineering undergraduate subjects.

Esther Titilayo Akinlabi Professor Esther Titilayo Akinlabi is a Full Professor at the Department of Mechanical Engineering Science, Faculty of Engineering and the Built Environment, University of Johannesburg, South Africa. She has had the privilege to serve as a Head of the Department of Mechanical Engineering Science and as the Vice Dean for Teaching and Learning at the University of Johannesburg. Her research interest is in the field of modern and advanced manufacturing processes—friction stir welding and additive manufacturing. Her research in the field of laser-based additive manufacturing includes laser material processing and surface engineering. She also conducts research in the field of renewable energy, and biogas production from waste. She is a rated National Research Foundation (NRF) researcher and has demonstrated excellence in all fields of endeavours. Her leadership, mentorship, and research experience are enviable as she guides her team of postgraduate students through the research journey. She is a recipient of several research grants and has received many awards of recognition to her credit. She is a member of the prestigious South African Young Academy of Science and is registered with the Engineering Council of South Africa. Professor Akinlabi has filed two patents, edited two books, published five books, and authored/co-authored over 400 peer-reviewed publications.

Chapter 1
Basics of Fused Deposition Modelling (FDM)

Abstract In this chapter, an overview of the basic principles of fused deposition modelling, commonly known as 3D printing technology, is presented. The chapter begins by introducing the holistic concept of additive manufacturing and its scientific principle as the technology for the modern and future industry. Then, the science of 3D printing is described. The applications of FDM in various fields are also highlighted with a focus on an interesting role the 3D printing technology is playing in the fight against Covid-19 pandemic. The chapter also gives a highlight of the parameters involved in fused deposition modelling of polymers and their basic interaction with the properties of the manufactured components. In relation to the process parameters, quality aspects of FDM products have also been briefly described in the chapter.

Keywords Additive manufacturing · Defects, fused deposition modelling · Quality of prints · Surface roughness

1.1 Additive Manufacturing

Additive Manufacturing (AM) involves classes of manufacturing technologies which build 3D components by adding a material layer upon a layer. The material could be a polymer, concrete, metal or even a composite. For a manufacturing process to qualify to be classified as an AM technique, it must involve the following three significant aspects.

- The use of a computer and computer aided design (CAD) to create visual 3D models: There are several CAD tools that are used to generate 3D models some of which include AutoCAD, Inventor®, Solidworks®, CATIA™ and so many others. Some of these software are available open source or closed source [1]. The technologist or engineer involved in the field of additive manufacturing should understand how to use a few or many of the software for effective manufacturing through these technologies. Through these CAD tools, and based on the experience of the user, any form of complex 3D models of the products can be generated. The amount of material to be extruded by the 3D printer and the time it will take

© The Author(s), under exclusive license to Springer Nature Switzerland AG 2020   1
F. M. Mwema and E. T. Akinlabi, *Fused Deposition Modeling*,
Manufacturing and Surface Engineering,
https://doi.org/10.1007/978-3-030-48259-6_1

to build the 3D model is determined and the information is created in a G-code file, which the printer can easily interpret [2].

- Slicing and generation of tool paths: The CAD 3D-generated models must be prepared in a format which can be interpreted by the additive manufacturing machine. The slicing software transforms the 3D design into layered models which the machine tool can easily trace. There are so many slicing software in the market and they are provided under different trademark names such as Cura, PrusaSlicer, MatterControl, Simplify3D, Repetier, ideaMaker, Z-SUITE, Slic3r, IceSL, Slicer-Crafter, Astroprint, 3DPrinterOS, SelfCAD, KISSlicer, Tinkerine Suite, Netfabb Standard including others and each of the software operates differently to achieve the best slicing [3].
- Conversion of the 3D model into real product: An additive manufacturing machine such as 3D printer and laser convert the 3D model into an actual product using engineering materials such as plastics, metal powders, composites, among others. The material(s) is melted and then allowed to flow according to the G-code (tool path) from the slicing software to create the 3D component.

There are various additive manufacturing methods, classified according to the material and machine technology used in the production of the components. According to the American Society for Testing and Materials (ASTM F42-) standards of 2010, there are seven categories of AM processes [4, 5] as listed below.

   i.   Material extrusion techniques
  ii.   Powder bed fusion techniques
 iii.   VAT photopolymerization methods
 iv.   Material jetting techniques
  v.   Binder jetting techniques
 vi.   Sheet lamination techniques
vii.   Direct energy deposition techniques.

The above processes utilize different materials and machines to create 3D printed components and have been extensively reviewed in the literature [6, 7]. Additive manufacturing processes are preferred over conventional processes due to the following advantages.

   i.   Enhanced material efficiency since no material wastage through cutting or machining.
  ii.   There is higher efficiency in resources since these processes do not require auxiliary resources such as tools, jigs, fixtures and so forth.
 iii.   Products of high complexity and intricacy can be manufactured since there are no constraints of the tools.
 iv.   Additive manufacturing processes enhance production flexibility.

Although these processes are attractive, they are constrained by some limitations such as size of parts that can be manufactured, surface and microstructural imperfections, and high cost of the AM equipment [7]. The processes are also very slow and therefore they are challenging technologies in mass production.

The focus of this book is on the strategies for improving the quality of the fused deposition modelling (FDM) of products. The reason for singling out FDM from all the many AM manufacturing processes is due to its wide range of applications and adoption by many individuals and industries. The FDM process, which is classified as the material extrusion AM technique, is the simplest, affordable and readily available 3D printing technique for polymer-based materials and it has been extensively used in various industries [8–11]. In the subsequent subtopics, the basics of FDM and its applications, parameters and quality aspects of the process are highlighted.

1.2 Science of FDM and Applications

Fused deposition modelling (FDM), also known as the material extrusion additive manufacturing technique, utilizes polymers as the raw material (filament). The filament is usually heated to a molten state and then extruded through the nozzle of the machine (3D printer). The nozzle head can move in three degrees of freedom (DoF) to deposit the extruded polymer on the build plate as per the G-code instructions. The principle of the FDM process is illustrated in a schematic diagram in Fig. 1.1. As shown, the filament is continuously fed through the extruder and nozzle of the machine via the two rollers rotating in opposite directions. The material is deposited on the build plate layer-by-layer until the required product shape and size are achieved. During the layering, the printer nozzle navigates back and forth as per the spatial coordinates of the original CAD model in the G-code files until the desired size and shape of the component is produced. In some FDM systems (3D printers), multiple extrusion nozzles can be used to deposit the polymer constituents especially in cases where components of compositional gradients are required. Usually, the resolution and effectiveness of the extrusion largely depend on the properties of the thermoplastic filament and as such, different 3D printers are designed for specific filament materials. In fact, most of the low-cost FDM 3D printers can process only one type of thermoplastic and polylactic acid (PLA) is the most common material. The components are usually layered onto the build plate (platform), which after printing can be removed by snapping off or soaking in a detergent depending on the type of the thermoplastic. Then, the printed components may be surface cleaned, sanded, painted or milled to enhance both their surface appearance and functionality.

There are various materials used in FDM and as stated earlier, PLA is the most adopted material by most 3D printer users at domestic and industrial levels due to the following reasons:

i. Polylactic acid (PLA) is a bioplastic and therefore eco-friendly and not harmful to human and animal health. PLA is a green material since it is fabricated from fully renewable sources such as corn, sugarcane, wheat or any other high carbohydrate containing resources [12]. As such, it is recommended for use in making cooldrink cups, deli and food take aways, and packaging containers.

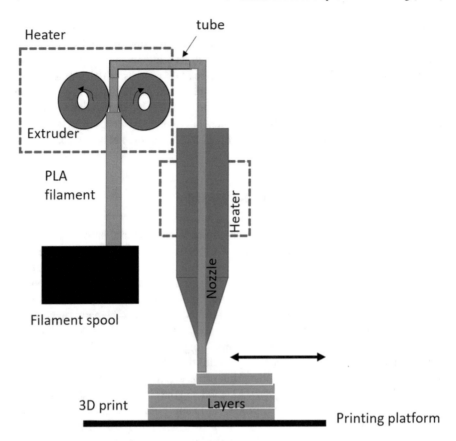

Fig. 1.1 Principle of fused deposition modelling

ii. PLA has a glass transition temperature ranging between 50 and 70°C and a melting point temperature ranging between 180 and 220°C [13–15]. As such, most low-energy and cost-effective 3D printers can extrude it. It is harder than Acrylonitrile butadiene styrene (ABS) although it (PLA) has higher friction when compared to ABS and therefore susceptible to extrusion blockage.

iii. PLA plastics are compostable and break down quickly upon disposal unlike the other plastics, which have posed serious disposal challenges. Being among the biopolymers, PLA degrades to natural and non-poisonous gases, water, biomass and inorganic salts when it is exposed to natural conditions, hydrolysis or even when incinerated.

iv. In its semi-crystalline form, PLA has shown to exhibit good flexural modulus, better tensility and flexural strengths.

v.   PLA is preferred by most 3D printer users because it does not always need a heated bed for the adhesion to occur between the print and the platform. Graphene-doped PLA, however, presents a great challenge for non-heated bed printers and it does not produce quality prints on non-heated build plates.

vi. PLA is commercially available in the market in a variety of colours and textures. This makes it attractive for users, especially domestic and decorative 3D printer handlers. The availability in various colours and texture has expanded the markets for CAD designers and toy enthusiasts. As such, the designers can develop interesting ideas and post in various databases (such as TurboSquid, CG Trader, Shapeways, Cults3D, 3DSquirrel and Thingsverse) where the toy enthusiasts can purchase, download and print with a variety of colours and texture designs of the PLA filaments.

Other materials used in FDM processing include polycaprolactone (PCL), polypropylene (PP), polyethylene (PE), polybutylene terephthalate (PBT), Acrylonitrile butadiene styrene (ABS), wood, nylon, metals, carbon fibre, graphene-doped PLA, etc. [13, 16, 17]. These materials are available in different commercial brands and trademarks, as filament wires, and can be purchased through various online stores such as Alibaba, Amazon and so forth. However, it is advised that the buyers should be aware of the chemical composition of the filaments they would like to use based on their applications. From the experience of the authors of this book, most of the filament suppliers do not provide reliable information regarding the chemical constituents of the 3D printing filaments and it is therefore recommended for the users, if necessary, to conduct their analyses to confirm the chemistry of these materials. These analyses can be conducted through phase identification on microscopy, X-ray diffraction (XRD) or more advanced chemical analysis facilities at their disposal.

The most common applications of FDM in modern society are listed below.

i. The technology has emerged as one of the most progressive methods for producing prototypes and rapid tooling of complex products in low and medium batches [18]. The research currently is on the development of a larger pool of materials for rapid prototyping applications and a lot of literature is available on this subject [19].

ii. There is an increasing adoption of the FDM technique in the toy and other related industries either as a direct manufacturing method [20] or method for producing moulds for injection moulding for such industries [21].

iii. The potential of FDM on mass personalization of products cannot be overemphasized. Due to flexibility and capability to produce intricate profiles, FDM finds application in producing customized products for various applications, for instance, personalized toys, automobile parts, interior design components, implants, beauty products and so forth [20].

iv. The FDM is also being applied in the medical field to produce moulds for casting of implants, medical devices and implants. The most exciting application is the 3D printing of moulds for investment casting of medical implants [22]. In traditional investment casting, there is the use of metallic moulds and sacrificial patterns (e.g. wax) to create the complex shapes of any implant. Therefore, using 3D printed moulds eliminates the need of having to use the sacrificial material and hence reduces cost, time and material wastage. However, there are still challenges associated with the integration of FDM into the investment casting process, that is, poor surface quality; as such, as illustrated in the literature herein (for example, [23] and others) a lot of research is currently underway in improving the surface properties of 3D printed parts and castings obtained from FDM moulds.

v. Other applications of FDM include direct printing of electrochemical cells for energy storage devices [24], micro-trusses for biomedical scaffolding [25], drug delivery components in the pharmaceutical industry [26], direct printing of conductors for electronic industry [27] among others.

1.3   3D Printing and the Novel Coronavirus (Covid-19) Pandemic

The recent outbreak of the Novel Coronavirus (Covid-19) across the entire world has led to a serious shortage of medical supplies and protective gears. To curb the spread of the Covid-19 pandemic, the World Health Organization (WHO) issued the following guidelines in February 2020:

i.   Restricted movement of people across countries.
ii.   Maintenance of high level of personal hygiene through handwashing with soap and the use of alcohol-based sanitizers.
iii. Use of protective gears for mouth, eyes and nose since the virus enters the human body through these membranes.
iv.   Avoidance of direct contacts among individuals and as such, individuals should keep at least 1-metre distance among themselves, and individuals should avoid handshakes, kissing and hugging.
v.   Avoiding direct touching/contact of surfaces as much as possible since the virus can survive on the surfaces for around 12 h.

These guidelines present a new challenge to both medical and science/engineering fields. As of the writing of this manuscript, the number of infections across the world was more than 1.69 million people with more than 102,000 deaths (www.worldo meters.info/coronavirus/). The number seems to be growing exponentially and this has considerably strained the health and medical sectors in terms of equipment and human capacity. There is an increasing demand for personal protective gear for both the public and medical practitioners. There is also a major shortage of medical ventilators and oxygen valves across the world and with the ban of international flights and travels, consumer-based regions such as Africa must innovate on availing these facilities. With several industries and technologies mobilizing their resources to contribute to this course, the 3D printing community has also been involved in the following ways:

i.   Availing open-source designs for medical and protective devices: Several indi-viduals and companies have developed designs for face mask frames, oxygen valves, nasal swabs, hands-free door openers and so forth, and availed them for free through the social media platforms and dedicated CAD online forums.

ii. Mass production of face masks: Most 3D printing companies such as Stratasys® and Prusa Research (Czech) have mobilized their global resources to undertake mass production of masks for both medical practitioners and general public. Individual 3D desktop printer owners across the world have also volunteered to print face masks for medical doctors. For example, in South Africa, a company known as 3D Printing Factor (PTY) Limited located in Johannesburg mobilized all her resources and individual companies to produce face shield frames for medical practitioners for free in partnership with Netcare 911 hospital (www.timeslive.co.za/news).

iii. Production of hands-free door openers: Several companies have 3D printed arm door openers to prevent people from touching the door handles directly. BCN3D Inc. based in Barcelona is one of those companies (www.3dprintingindustry.com).

iv. Manufacturing of medical ventilator parts: ISINNOVA Limited, a company based in Brescia (https://www.isinnova.it/), Italy, has 3D printed oxygen valves to assist patients exhibiting respiratory difficulties. A consortium of companies in collaboration with Zona Franca Consortium (CZFB) and Leitat Technology Centre (Spain) have developed and tested a 3D printed respiration equipment for the Covid-19 patients. The automotive giant, Volkswagen (Germany), has also invested in 3D printing of ventilators for hospitals. Photocentric Group Inc., UK, has produced more than 600 test units of respiratory valves for the patients of Covid-19 (https://photocentricgroup.com/).

v. Fabrication of quarantine facilities: In China, Winsun Limited has 3D printed several quarantine rooms for Xianning Central Hospital in Wuhan. The walls of the 3D printed houses have been shown to be stronger than the conventional walls (www.3dprintingindustry.com).

vi. Production of testing kits: Some 3D printing companies have designed and manufactured nasal swabs to quicken mass testing for the virus across the different parts of the world. An example of such companies is Formlabs, a 3D printing company based in Massachusetts, US (https://formlabs.com/). The company has the capacity to produce more than 100,000 nasal swabs a day. This is going to enhance mass testing for the virus across different parts of the world.

vii. 3D printing also has the capacity to develop intricate moulds for injection moulding of parts for medical uses by the hospitals to help the Covid-19 patients.

viii. The was 3D printing of drones for delivery of groceries and medical supplies in China during the lockdown periods (www.dezeen.com). Also, the use of 3D printing of drones has been undertaken in South Africa by the Centre for Scientific and Industrial Research (CSIR) (www.sanews.gov.za/south-africa).

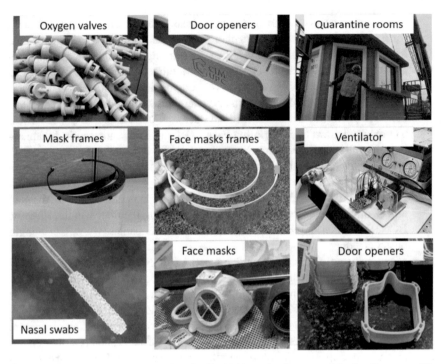

Fig. 1.2 Some of the 3D printed devices for Covid-19 pandemic accessed for free from www.3dp rintingindustry.com

Figure 1.2 shows some of the 3D printed components which have been designed and manufactured, so far, to help in fighting against the spread of the Covid-19 virus as well as assisting its patients.

1.4 Process Parameters in FDM

Fused deposition modelling (FDM) is influenced by various parameters as summarized in Fig. 1.3. As shown, the parameters have been classified into two broad categories, namely machine and material parameters. The machine parameters are those parameters the 3D printer user will specify on the slicing software during the generation of the G-code files whereas the material parameters are the properties of the filament material or materials being extruded through the nozzle. Some of the machine parameters, as shown, include the printing speed, raster angle, melt flow rate through the nozzle, airgap, layer thickness, infill density, build orientation and temperature [10, 11]. On the other hand, the material properties such as thermal and mechanical influence both the extrusion and performance of the print.

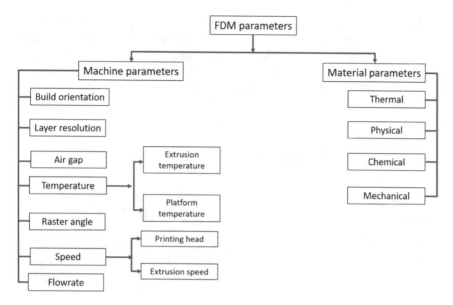

Fig. 1.3 Parameters influencing the fused deposition modelling process

The quality and performance of the printed parts depend on the choice of these parameters and there are various efforts in the literature on evaluating the effects of various parameters on the process and quality of the prints [28]. The build orientation basically indicates the angle at which the longest length is inclined to the base of the build plate. The printed components may be inclined at $0°$, $45°$, $90°$, etc. depending on the choice of the user. Layer resolution indicates the minimum thickness of every layer in one run of the print head and it may vary from a few micrometres to millimetres depending on the accuracy and application of the 3D printer. The extrusion temperature measures the temperature supplied from the external source to the printer heating elements to melt the filament material for easy extrusion whereas the platform temperature is the temperature applied on the build plate to enhance the adhesion of the prints onto the platform and avoid printing failure.

It is noted that not all printers have a heated bed and except PLA, all the other materials require heating for them to stick onto the platform. At times, when using non-heated bed with some materials, it has become a common practice to use some sticking fluids such as office glue to enhance the sticking of the first layer of the print. However, from the experience of the authors on 3D printers, it usually affects the dimensional accuracy of the print and gradually blocks the nozzle orifice. Blockage of the nozzle passage may lead to a major failure of the 3D printer and replacement of some parts. During printing, the extrusion temperature should be set within the melting point of the filament material. The thermal properties of the material will influence the conditions of melting and flow through the nozzle of the 3D printer. The chemical properties of the filament material determine the glass transition region and hence the quality of the printed part. Mechanical properties such as strength and friction in molten state determine the rigidity of the printed part and the flowability of the material and whether the material will jam the nozzle or not. The choice, optimization and interrelationships among the process parameters to the print quality are the main objectives of this book and will be detailed using literature case studies and some results obtained from research conducted by the authors in the subsequent chapters.

1.5   Quality Issues in FDM

As mentioned, FDM involves the layering of molten filament material to create the desired product. The adhesion and fusion between adjacent layers are very critical for quality prints. Additionally, the extrusion conditions of the filament material during the printing process affect the accuracy, quality and performance characteristics of the printed product. The surface roughness of the FDM products is one of the major drawbacks of the quality of this process. Due to the nature of the process, the surfaces of the product mostly exhibit the 'back-and-forth' tracks of the printing nozzle known as the stair stepping effect (shown in Fig. 1.4). These tracks create terraces on the surface, therefore, leading to relatively high average roughness values (Ra) in the range of micrometres. Such high levels of roughness ranges impede the application of the FDM manufactured products in some fields such as dentistry, biomedical, sensing and so many other areas of applications. The presence of terraces and deeps on the surface of the prints can lead to penetration of moisture and other environmental electrolytes into the inner layers of the product causing further degradation of its properties. For instance, such components (with high roughness) would be very detrimental for use as prosthesis or dental implants as they would react with body fluids, which would cause premature failure of the implants. At times, due to improper extrusion (over- or under-extrusion), there is a lack of enough adhesion between the adjacent layers of the filament material that enhances high roughness and other structural defects such as porosity and cracks.

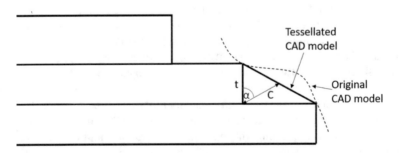

Fig. 1.4 Illustrating the staircasing effect of the FDM parts. In this case, C is known as cusp height, t is the layer thickness and α is the angle between the cusp height and layer height

The surface roughness of 3D printed parts is predicted using optical surface profilers and microscopes, and various post-processing methods have been adopted to lower the surface roughness. These processes are either mechanical or chemical methods. The most commonly used mechanical methods for enhancing surface quality of FDM prints are machining, sanding, polishing, abrasion and barrel finishing whereas the chemical methods involve painting, coating, heating and vapour deposition [19]. It has been researched and reported in the literature that the choice of each of the methods depends on the materials and performance requirements of the FDM manufactured part. Additionally, at the design stage of the components, the stair stepping effects of the printer can be minimized by optimizing the slicing procedure and print resolution. Using very large slicing thickness reduces the printing time and produces very rough products due to the stair stepping effect. On the contrary, fine slicing reduces the stair stepping effect and reduces the surface roughness although it results in longer printing times, which may impact the other aspects of manufacturing, especially during mass production. The different strategies for slicing have also been shown to influence the quality of the print [29, 30].

The lack of adhesion leading to structural defects considerably affects the dimensional accuracy and mechanical integrity of the FDM printed components. It is obvious that components consisting of a very high density of defects would experience dimensional errors and low properties such as hardness, flexural strength, tensility and compression, and impact strengths. If there is not enough adhesion between the layers, the filament material of the adjacent layers will be forced to flow and compensate between the resulting spaces. This may lead to shrinkage of the component causing dimensional errors between the CAD design and the actual print. Additionally, the presence of pores and cracks within the structure increases the stress raisers within the material such that the component cannot absorb the required energy during its performance without failure. These defects further enhance the propagation of the cracks and the components may not offer suitable and enough mechanical stability for various applications.

The flow rate of the filament material during the extrusion and deposition also plays an important role on the quality of the 3D printed parts. The choice of the extrusion and heated bed temperatures are based on the flow characteristics of the filament material. Insufficient flow of the molten material between the layers causes spaces or incomplete adhesion; these weaken the component and failure can easily occur through delamination. The raster angle is also related to the material flow during the FDM process; for instance, Galeja et al. [31] have recently published an article in *Materials (Basel)* journal titled 'static and dynamic mechanical properties of 3D printed ABS, a function of raster angle.' The study demonstrated that for a range of raster angles 45°–90°, the raster angle of 55° provided the optimal flow of molten ABS during printing and at that angle, the printed ABS samples exhibited excellent static and dynamic responses to mechanical loads.

There is a continued effort by the scientific community to understand the influence of the specific parameters to the FDM process and the quality of the printed product. As illustrated in the previous section, the interactions among these parameters in an FDM process are complex and require multi-objective approaches to understand and enhance the quality of manufacturing. The general approach in every manufacturing process including FDM is to understand the influence of the individual parameters followed by evaluating the most significant of those factors and finally determining combined effects (multi-objective) of the parameters. There are several publications describing both of these approaches for enhancing quality in FDM processes [11, 22, 32–34]. For instance, Perez et al. [35] investigated the effect of five FDM printing parameters (speed, wall thickness, layer height, temperature and printing path) on the surface roughness of PLA printed components using analysis of variance (ANOVA), graphical methods and non-parametric tests. It was reported that wall thickness and layer height were the most significant factors for surface roughness as compared to the other factors. It has been reported that the optimal choice of the wall thickness and enhancement of the geometry of the product (by defining clearly the G-code) eliminates the microstructural defects in FDM PLA products [34]. In another study, Singh et al. [23] optimized the quality of the FDM printed ABS samples for prosthetic investment casting using Taguchi L_{18} orthogonal array for the print orientation, fill density, vapour smoothing parameters and heat treatment time of the samples.

The choice of the optimal conditions or settings for quality FDM printing during large-scale and mass customization production becomes more difficult. In such cases, the manufacturer is posed with the challenge of quality and time of manufacturing. Usually, 3D printing is a very slow process and most of the Desktop 3D printers are slower compared to the other manufacturing processes. For instance, it is discussed in the literature that the stair-step effects of the FDM process can be eliminated by using very small layer height. Thin layers mean that a very small volume of the material is layered for every run as compared to relatively larger layer thicknesses. It means that longer time is required for enough melting and flow of the filament material within the layers to eliminate microstructural defects such as porosity and cracks. During FDM printing, the process should be closely monitored, especially during the initial stages and the following simple observations are recommended to ensure geometrically and dimensionally accurate prints:

- When the filament starts loading, remove the molten filament forming around the nozzle to avoid clogging and blockage. This filament is usually under-molten and cannot fuse strongly with the rest of the print material.
- The initial point of printing during the creation of the base support structures (e.g. brim and raster) should be cut off to avoid it being dragged by the nozzle and then destroying the entire support structure.
- Observe the adhesion of the support structure onto the build plate and if the structure appears to delaminate from the plate, stop the printer and repeat the printing process. Additionally, if the printer does not have a heated bed, and the PLA filament does not stick, office glue or other types of glues can be applied onto the printing table to enhance the adhesion and hence, the quality of the prints.
- Observe the critical points of the print (corners, holes, etc.) and check for any incomplete fusion, lack of proper filling and gaps between the layers. These deformities are caused by over-extrusion or under-extrusion of the filament and when this happens, the printer should be stopped, and the settings adjusted.

- The occurrence of stringing or formation of strings/hairs especially when the printer is moving between different sections of the print should also be checked.
- The layering consistency of the print should be checked; slight shifting or separation of the layers is an indication of poor-quality printing.
- The warping of large parts during the FDM processing causes poor sticking of the printed component onto the build plate such that the component shifts on the build plate causing form and dimensional errors.

In Chap. 4 of this monograph, some of these quality challenges related to surface quality are illustrated. The challenges can greatly slow the manufacturing process through fused deposition modelling since they imply frequent switching ON and OFF of the printer which prolongs the printing time and results in material loss. It is common knowledge that the longer the manufacturing time, especially in mass production, the higher the cost of production and hence, the process becomes economically unviable.

1.6 Summary

The basics of fused deposition modelling have been discussed in the chapter. It is no doubt that technology has significantly been accepted in the direct manufacturing of components besides being an attractive rapid prototyping method. It is being utilized to manufacture biomedical devices during the Novel Coronavirus (Covid-19) pandemic. The quality of the FDM parts depends on the manufacturing process settings. There are several parameters directly influencing the quality of the FDM parts some of which include temperature, speed, infill density, layer height and build orientation. Improper choice of the parameters may lead to adhesion problems between the layers, therefore causing the formation of defects. The major drawback of FDM parts is high surface roughness due to the stair-stepping effect of a 3D printer.

The goal of any 3D printer user is to produce high-quality products in terms of form and dimensional accuracy. In the subsequent chapters, the strategies for enhancing the quality of FDM parts based on experiments conducted by the authors and other peer-reviewed and published data will be presented.

References

1. S. Junk, C. Kuen, Review of open source and freeware CAD systems for use with 3D-printing. Proc. CIRP **50**, 430–435 (2016)
2. Y. Song, Z. Yang, Y. Liu, J. Deng, Function representation based slicer for 3D printing. Comput. Aided Geom. Des. **62**, 276–293 (2018)
3. B. Huang, S.B. Singamneni, Curved layer adaptive slicing (CLAS) for fused deposition modelling. Rapid Prototyp. J. **21**(4), 354–367 (2015)
4. A. Lele, Additive manufacturing (AM), in *Smart Innovation, Systems and Technologies*, vol. 132 (Springer, Berlin, 2019), pp. 101–109

5. A.M. Forster, Materials testing standards for additive manufacturing of polymer materials: State of the art and standards applicability, in *Additive Manufacturing Materials: Standards, Testing and Applicability*, pp. 67–123 (2015)
6. W.S.W. Harun et al., A review of powdered additive manufacturing techniques for Ti-6al-4v biomedical applications. Powder Technol. **331**, 74–97 (2018)
7. K.S. Prakash, T. Nancharaih, V.V.S. Rao, Additive manufacturing techniques in manufacturing—An overview. Mater. Today Proc. **5**(2), 3873–3882 (2018)
8. M.A. Cuiffo, J. Snyder, A.M. Elliott, N. Romero, S. Kannan, G.P. Halada, Impact of the fused deposition (FDM) printing process on polylactic acid (PLA) chemistry and structure. Appl. Sci. **7**(6), 1–14 (2017)
9. W.C. Lee, C.C. Wei, S.C. Chung, Development of a hybrid rapid prototyping system using low-cost fused deposition modeling and five-axis machining. J. Mater. Process. Technol. **214**(11), 2366–2374 (2014)
10. A.D. Valino, J.R.C. Dizon, A.H. Espera, Q. Chen, J. Messman, R.C. Advincula, Advances in 3D printing of thermoplastic polymer composites and nanocomposites. Prog. Polym. Sci. **98**, 101162 (2019)
11. A. Dey, N. Yodo, A systematic survey of FDM process parameter optimization and their influence on part characteristics. J. Manuf. Mater. Process. **3**(3), 64 (2019)
12. J.C. Camargo, Á.R. Machado, E.C. Almeida, E.F.M.S. Silva, Mechanical properties of PLA-graphene filament for FDM 3D printing. Int. J. Adv. Manuf. Technol. **103**(5–8), 2423–2443 (2019)
13. Y. Liao et al., Effect of porosity and crystallinity on 3D printed PLA properties. Polymers (Basel) **11**(9), 1487 (2019)
14. A. Rodríguez-Panes, J. Claver, A.M. Camacho, The influence of manufacturing parameters on the mechanical behaviour of PLA and ABS pieces manufactured by FDM: A comparative analysis. Materials (Basel) **11**(8), 1333 (2018)
15. J. Kiendl, C. Gao, Controlling toughness and strength of FDM 3D-printed PLA components through the raster layup, *Compos. Part B Eng.* **180** (2020)
16. B. Mansfield, S. Torres, T. Yu, D. Wu, A review on additive manufacturing of ceramics, in *ASME 2019 14th International Manufacturing Science and Engineering Conference, MSEC 2019*, vol. 1, pp. 36–53 (2019)
17. J.R.C. Dizon, A.H. Espera, Q. Chen, R.C. Advincula, Mechanical characterization of 3D-printed polymers. Addit. Manuf. **20**, 44–67 (2018)
18. S. Singh, R. Singh, Integration of fused deposition modeling and vapor smoothing for biomedical applications, in *Reference Module in Materials Science and Materials Engineering*, Elsevier, pp. 1–15 (2017)
19. K.S. Boparai, R. Singh, Development of rapid tooling using fused deposition modeling, in *Additive Manufacturing of Emerging Materials* (Springer International Publishing, Cham, 2019), pp. 251–277
20. M.A. León-Cabezas, A. Martínez-García, F.J. Varela-Gandía, Innovative functionalized monofilaments for 3D printing using fused deposition modeling for the toy industry. Proc. Manuf. **13**, 738–745 (2017)
21. C. Whlean, C. Sheahan, Using additive manufacturing to produce injection moulds suitable for short series production. Proc. Manuf. **38**, 60–68 (2019)
22. D. Singh, R. Singh, K.S. Boparai, Development and surface improvement of FDM pattern based investment casting of biomedical implants: A state of art review. J. Manuf. Process. **31**, 80–95 (2018)
23. D. Singh, R. Singh, K.S. Boparai, I. Farina, L. Feo, A.K. Verma, In-vitro studies of SS 316 L biomedical implants prepared by FDM, vapor smoothing and investment casting. Compos. Part B Eng. **132**, 107–114 (2018)
24. P. Chang, H. Mei, S. Zhou, K.G. Dassios, L. Cheng, 3D printed electrochemical energy storage devices. J. Mater. Chem. A **7**(9), 4230–4258 (2019)
25. M.P. Nikolova, M.S. Chavali, Recent advances in biomaterials for 3D scaffolds: A review. Bioact. Mater. **4**(August), 271–292 (2019)

26. V. Linares, M. Casas, I. Caraballo, Printfills: 3D printed systems combining fused deposition modeling and injection volume filling. Application to colon-specific drug delivery. Eur. J. Pharm. Biopharm. **134**, 138–143 (2019)

27. C.L. Manzanares Palenzuela, F. Novotný, P. Krupička, Z. Sofer, M. Pumera, 3D-Printed Graphene/Polylactic acid electrodes promise high sensitivity in electroanalysis. Anal. Chem. **90** (9), 5753–5757 (2018)

28. J.S. Shim, J.-E. Kim, S.H. Jeong, Y.J. Choi, J.J. Ryu, Printing accuracy, mechanical properties, surface characteristics, and microbial adhesion of 3D-printed resins with various printing orientations. J. Prosthet. Dent. **121**, 1–8 (2019)

29. B. Huang, S. Singamneni, Alternate slicing and deposition strategies for fused deposition modelling of light curved parts. J. Achiev. Mater. Manuf. Eng. **55**(2), 511–517 (2012)

30. J. Podroužek, M. Marcon, K. Ninčević, R. Wan-Wendner, Bio-inspired 3D infill patterns for additive manufacturing and structural applications. Materials (Basel) **12**(3), 1–12 (2019)

31. M. Galeja, A. Hejna, P. Kosmela, A. Kulawik, Static and dynamic mechanical properties of 3D printed ABS as a function of raster angle. Materials (Basel) **13**(2), 297 (2020)

32. J.S. Chohan, R. Singh, K.S. Boparai, Parametric optimization of fused deposition modeling and vapour smoothing processes for surface finishing of biomedical implant replicas. Meas. J. Int. Meas. Confed. **94**, 602–613 (2016)

33. I. Dankar, A. Haddarah, F.E.L. Omar, F. Sepulcre, M. Pujolà, 3D printing technology: The new era for food customization and elaboration. Trends Food Sci. Technol. **75**, 231–242 (2018)

34. E.G. Gordeev, A.S. Galushko, V.P. Ananikov, Improvement of quality of 3D printed objects by elimination of microscopic structural defects in fused deposition modeling. PLoS One **13**(6), e0198370 (2018)

35. M. Pérez, G. Medina-Sánchez, A. García-Collado, M. Gupta, D. Carou, Surface quality enhancement of fused deposition modeling (FDM) printed samples based on the selection of critical printing parameters. Materials (Basel) **11**(8), 1382 (2018)

Chapter 2
Print Resolution and Orientation Strategy

Abstract In this chapter, a strategy for enhancing surface roughness and hardness based on the full factorial design of layer resolution and build orientation is illustrated. Layer resolution levels of 0.1 mm, 0.2 mm and 0.3 mm and build orientations of 0°, 15°, 30°, 45°, 60° and 90° were used to develop the full factorial design of experiments (DOE). Analysis of variance (ANOVA) was undertaken to determine the statistical significance of the factors to roughness and hardness properties of the printed parts. The mean interaction plots of the data were also used to study the interrelationships among the responses and the two printing parameters. The results revealed that layer resolution is the most significant parameter influencing the mean surface roughness of the PLA printed samples whereas build orientation closely influences the surface hardness as compared to the layer resolution although the ANOVA reveals that both parameters are statistically insignificant as far as hardness is concerned.

Keywords ANOVA · Build orientation · Fused deposition modelling · Hardness · Layer resolution · Roughness

2.1 Introduction

The resolution of the print layers and the build orientation are parameters which play an important role in the quality of the FDM manufactured products. The layer resolution can be directly related to the staircasing effect of the layers and surface roughness whereas the orientation angle of the print quality can be related to the fusion and arrangement of the layers, and mechanical strength of the FDM part. The combined effect of both parameters (layer resolution and build orientation) to the quality of the print is of interest and as such, there are several studies attempting to investigate these parameters. Experimentally, the layer resolution indicates the layer thickness for each run of the 3D printer nozzle over the print whereas the orientation refers to the angle under which the print is inclined to the horizontal axis of the build plate. These parameters are usually set during the generation of the tool paths (G-codes) of the FDM machine in the slicing software. Evidence in the literature shows that the proper choice of the combination of these two parameters can greatly influence the quality of the 3D printed products for various filament materials.

Printing orientation has been shown to influence the printing accuracy and mechanical strength of 3D printed products. For instance, a study by Shim et al. [1] investigated the relationship among the printing orientation to the accuracy, flexural strength, roughness and microbial behaviour of FDM manufactured resin for dental applications. The study utilized three orientations, namely, 0°, 45° and 90° to produce the denture base of Poly(methyl methacrylate) (PMMA) through the FDM technique. The study revealed that samples printed at 90° orientation exhibited the lowest error rates in dimensional accuracy while the 45° showed statistically higher error rates. The flexural strength was found highest in the prints produced at 90° while the lowest at 0°. It was also reported that the highest roughness was obtained on samples printed at 45° orientation. The microbial action was found highest on samples prepared at 90° while the lowest on samples printed at 0°. The study clearly illustrated the influence of print orientation on the 3D printed samples for dental applications.

In another study, [2] the failure strength and separation angle were investigated at different build orientations and layer thicknesses and it was shown that these parameters significantly affect the strength and failure characteristics of the 3D printed parts. The influence of the build orientation of polyacetal material (POM) on the strength of the 3D printing was reported [3]. Samples were printed at 0°, 45° and 90° and at ASTM D638 standard [3]. It was observed that the highest ultimate tensile strength was obtained at 0° build orientation. Similar results were reported for samples prepared at flat, width and upright orientations, in which the flat orientation was shown to have the highest impact and tensile properties [4]. A study by Afrose et al. [5] investigated the effect of the part orientation on the tensile fatigue properties of the PLA component. It was reported that parts oriented in the X-direction (flat) exhibited higher tensile stresses in static loading than those printed in Y- and 45°-orientations. However, in cyclic loading, samples printed in 45° orientation exhibited higher fatigue life as compared to parts printed in X- and Y-orientations. Kovan et al. [6] investigated the effect of layer thickness and print orientation on mechanical strength and adhesion bonding of 3D printed PLA samples. The study reported that edgewise orientation of the build part had the highest adhesion strength when printed at lower thicknesses whereas, on flat orientation, the highest adhesion strength was obtained at a higher thickness of the print layers. The study also reported a relationship between the adhesion strength and print surface roughness.

As a result of the interactions between print orientation and layer resolution in FDM technology, several researchers have embarked on optimization to determine the best combinations of these parameters for enhanced quality and performance of the printed parts. The influence of raster angle, infill density and layer height on the mechanical integrity of Acrylonitrile butadiene styrene (ABS) printed parts was investigated [7]. The interrelationship was determined using a response surface methodology in which the optimal printing parameters were determined as 80% infill, 0.5 mm layer thickness and 65° raster angle. In another study, layer thickness, build orientation, number of contours and infill density were optimized using advanced techniques namely response surface methodology-genetic algorithm (RSM-GA) and artificial neural network (ANN) [8]. The study demonstrated that these tools can sufficiently predict the indirect correlation between various FDM parameters and the quality of the prints. Grey relational analysis was applied to determine optimal layer thickness and print orientation for samples prepared from bronze PLA, wood PLA and TPU material filaments [9]. The surface roughness of FDM printed samples was optimized using RSM, particle swarm optimization and symbiotic organism search algorithms in relation to the print parameters (layer height, print speed, print temperature and outer shell speed) [10]. Mohamed et al. [11] used a fractional factorial design to determine the effect of six FDM parameters (layer thickness, air gap, raster angle, build orientations, road width and number of contours) on the dynamic mechanical properties of 3D printed parts. Using analysis of variance (ANOVA), the study revealed that the most influential parameters were the layer thickness, air gap and number of contours. The best combination for enhanced dynamic performance was 0.332 mm (layer height), air gap (0.00 mm), raster angle (0.0°), orientation angle (0.0°), road width of 0.4572 mm and 10 contours.

From the proceeding evidence in the literature, layer resolution and print orientation greatly influence the quality of the FDM printed parts. These parameters are straightforward to set and control in the FDM printers. For most 3D printers, the optimal layer resolution varies from a few micrometres to millimetres and since it greatly influences the production time, too low resolution implies longer printing time and vice versa. However, too high layer height may lead to high roughness and compromise on the adhesion strength between the layers. As such, a balance between manufacturing time (hence cost) and quality must be stricken. On the other hand, build orientation greatly affects both surface roughness and strength of the parts. However, some of the orientations may require additional support structures besides the raft/brim to avoid premature breakdown of the part during the printing process. Such structures imply more material usage and may greatly influence the overall cost of manufacturing especially for mass production. Here, a full factorial design is adopted to explain build orientation and layer resolution as strategies for quality enhancement in FDM printing of PLA parts.

2.2 Materials and Methods

This section presents the materials and methods employed in an experiment conducted by the authors on illustrating the use of the design of experiment for FDM parameters (build orientation and layer resolution) as a strategy for quality enhancement for PLA printed parts.

In this experiment, print resolution and build orientation were considered as the process parameters. The printing resolution was considered under three levels whereas there were six levels of the build orientation as shown in Table 2.1. The lower limit of the print resolution is 0.05 mm and the upper limit is 1.2 mm and the recommended range of the resolution is between 0.1 mm and 0.3 mm. The build

Table 2.1 3D printing process parameters and their levels

Parameter	Levels					
	1	2	3	4	5	6
Printing resolution (mm)	0.1	0.2	0.3	–	–	–
Build orientation (°)	0	15	30	45	60	90

orientation levels are chosen based on previous studies [2] and the experience of the authors on FDM printing.

Based on the process parameters and levels shown in Table 2.1, the general full factorial ($3^1 \times 6^1$) was used to design the experiment, and the experimental matrix is generated as shown in Table 2.2.

The other parameters shown in Table 2.3 were kept constant throughout the experiments. These parameters were chosen based on the experience of the authors with their 3D desktop printer.

The test results of surface roughness (Ra) and Rockwell hardness (HR) of the fused deposition modelling (FDM) samples were used as the responses to the factorial design and signal-to-noise ratio (SNR). The analysis was undertaken in Minitab 17 software to determine the best combination of parameters for roughness and hardness characteristics of the FDM parts.

Table 2.2 Experimental matrix based on general full factorial design L_{18}

Exp. number	Level of printing resolution	Level of build orientation
1	0.1	30
2	0.2	15
3	0.1	60
4	0.1	45
5	0.1	15
6	0.2	30
7	0.2	60
8	0.2	45
9	0.3	45
10	0.1	90
11	0.2	90
12	0.3	60
13	0.3	0
14	0.2	0
15	0.1	0
16	0.3	15
17	0.3	30
18	0.3	90

Table 2.3 The constant 3D printing parameters used in the experiment

Parameter	Value/description
Shell thickness	0.8
Bottom/top thickness	0.6
Fill density	100%
Speed	50 mm/s
Temperature	210 °C
Support type	Touching the build plate
Adhesion type	Raft
Diameter of filament	1.75 mm
Travel speed	80 mm/s

The samples were prepared using a home-friendly WANHAO 3D printer, Duplicator 10 using the parameters in Tables 2.2 and 2.3. This printer has a non-heated bed and a maximum extrusion temperature of 210 °C and is suitable for PLA filament only [12]. The samples were of rectangular cross-sections of 10 mm × 5 mm × 15 mm illustrating the simplest models the 3D printer can manufacture. The models were designed in SpaceClaim® CAD software (ANSYS student version 2019) and exported as. *STL* file to Cura slicing software. The 3D printing settings (according to Tables 2.2 and 2.3) were set in the slicing software. The G-codes were generated for each of the experiments in Table 2.2 meaning eighteen (18) files were generated and saved in the memory card for the 3D printer. Figure 2.1 illustrates the build orientation variation during the 3D printing of the samples. The properties of PLA material used in this experiment are summarized in Table 2.4. The PLA filament was purchased from 3D Printing Factory (PTY) LTD (Boksburg store in Johannesburg, South Africa) and it is manufactured by WANHAO Ltd.

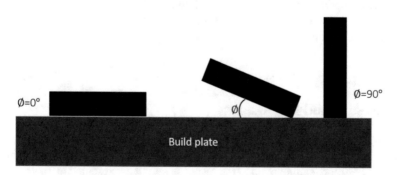

Fig. 2.1 Sample preparations at varying print orientations (∅). The samples were prepared at angles 0°, 15°, 30°, 45°, 60° and 90° at print resolutions of 0.1, 0.2 and 0.3. Such angles have been shown in the literature [2] to influence the surface quality and strength of PLA printed samples

Table 2.4 The constant 3D printing parameters used in the experiment

Property	Description
Filament diameter	1.75 ± 0.05 mm
Recommended printing temperature	190–230 °C
Density at room temperature	2.14 g/cm^3
Recommended speed of printing	50 mm/s
Softening temperature	146–150 °C
Printing resolution	0.1–1.2 mm
Temperature of platform	Non-heated (room temperature)

The 3D printed samples were then characterized for surface roughness (Ra) using handheld roughness tester (TR200; Time Group Inc.) and strength using a Rockwell microhardness machine. For each surface, ten values of both hardness and roughness were obtained with a cut-off of 0.25 mm, and averages obtained for statistical analyses. The average surface roughness (Ra) values were used in this study rather than the root mean square values since the objective was to evaluate the protrusions of the surface height features beyond the arbitrary datum. Additionally, most of the existing studies in the literature have utilized the average surface roughness parameter (Ra) rather than the root mean square to evaluate the quality of 3D printed samples. It is important to note that the values obtained were in micrometre scales indicating the significance of the roughness in the quality of PLA FDM products. The handheld roughness tester has also been utilized in various advanced studies of analysing machined and 3D printed surfaces [13] and it is preferred for micro-roughness analyses over atomic force microscopy (AFM) since AFM gives nano-roughness values and it is best suitable for thin-film depositions and characterizations such as sputtering, atomic layer deposition (ALD) and chemical vapour depositions [14, 15, 16, 17, 18]. However, it should also be noted that some studies on laser 3D printing of metallic components (laser cladding) have utilized AFM to characterize the nano-roughness properties of such processes, although with lots of challenges as can be observed in the study by Erinosho et al. [19]. In this case, the handheld roughness tester was scanned along at a scanning length of 10 mm for each roughness measurement.

2.3   Results and Discussions

The results of the characterization of the samples prepared at different process parameters and levels are summarized in the response output as presented in Table 2.5. As shown, experiment no. 9 did not show any results since our printer could not print at

Table 2.5 Response output table

Exp. number	Level of printing resolution (mm)	Level of build orientation (°)	Surface roughness (Ra), μm	Rockwell hardness
1	0.1	30	3.61	124.8
2	0.2	15	6.88	126.9
3	0.1	60	2.89	126.1
4	0.1	45	4.04	128.3
5	0.1	15	5.07	127.5
6	0.2	30	8.80	126.4
7	0.2	60	8.77	127.5
8	0.2	45	4.35	129.4
9	0.3	45	–	–
10	0.1	90	1.33	127.0
11	0.2	90	2.89	124.3
12	0.3	60	8.44	128.9
13	0.3	0	4.78	127.0
14	0.2	0	2.48	130.7
15	0.1	0	1.51	128.6
16	0.3	15	6.76	130.6
17	0.3	30	9.08	125.5
18	0.3	90	8.84	124.7

that parameter combinations. Various trials were undertaken at no success since the sample would delaminate from the build plate after forming the raft.

Analysis of variance (ANOVA) of surface roughness

The validation of the design of the experiment and assumptions in randomization of the tests are illustrated by the normality, constant variance and residuals versus order plots in Figs. 2.2, 2.3 and 2.4, respectively, for the average roughness (Ra) as the response. As shown, the distribution of the residuals is normal since the probability plots can be estimated by a straight line. The constant variance (plots of residuals against the fitted values) does not indicate any pattern and therefore implies that the assumption of constant variance is satisfied. The plot of residuals versus the time order of the experiments (observations) also does not show any order, further indicating the satisfaction of independence assumption.

Table 2.6 shows the one-way analysis of variance (ANOVA) results for surface roughness response such as P-value, coefficient of determination (R^2), standard deviation (S), adjusted R^2 and predicted R^2. The parameters are significant if the values are less than the alpha value, which is 0.05. As shown, the P-values were obtained as 0.098 and 0.007 for build orientation and resolution, respectively. It means that print resolution (layer height) according to this analysis greatly influences the surface

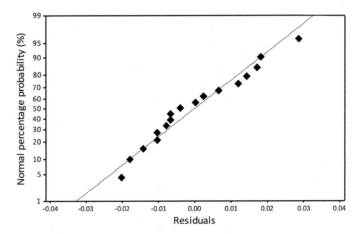

Fig. 2.2 Normality plots of the residuals for average roughness

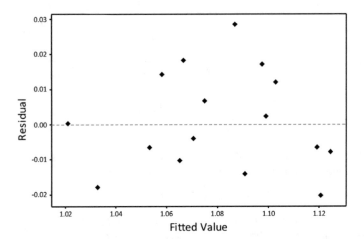

Fig. 2.3 Residuals versus fitted values (constant variance) plots

roughness as compared to the build orientation. As such, based on the ANOVA analysis, the general relationship among the specific parameters can be written as

Roughness = 5.272 − 2.349 Orientation(0°) + 0.964 Orientation(15°) + 1.891 Orientation(30°) − 1.016 Orientation(45°) + 1.428 Orientation(60°) − 0.919 Orientation(90°) − 2.197 resolution(0.1°) + 0.423 resolution(0.2°) + 1.774 resolution(0.3°)

As shown in Fig. 2.5, it was observed that the surface roughness increased proportionally with the resolution height. Furthermore, the ANOVA analysis showed that for a specific parameter influence of roughness, the only significant build orientation on the surface roughness was at 0° (P = 0.030) with the rest of the orientation angles showing values of P more than 0.05. Figures 2.6 and 2.7 show the means plots for roughness and surface plots in 2D and 3D, respectively. It can be seen from these

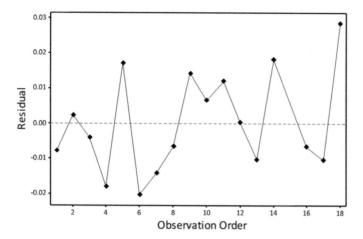

Fig. 2.4 Residuals versus experimental observation order

Table 2.6 One-way analysis of variance (ANOVA) results for surface roughness response

Source	DF	Adj SS	Adj MS	F-value	P-value
Model	7	94.81	13.54	4.66	0.018
Linear	7	94.81	13.54	4.66	0.018
Orientation	5	38.19	7.64	2.63	0.098
resolution	2	53.82	26.91	9.26	0.007
Error	9	26.14	2.91		
Total	16	120.95			
	Model summary				
	Standard deviation	R^2	R^2(adj)	R^2(pred)	
	1.70	78.39%	61.58%	21.68%	

plots that the best combination for low roughness is at resolution of 0.1 mm and build orientation of 0°. The plots in Fig. 2.6 show that there exists a main effect response, that is, the mean surface roughness (Ra) responses are not the same across all the levels. As further observed, the 30° orientation is associated with the highest mean Ra and 0° with the lowest mean Ra. However, as described above, the factor is not statistically significant to the roughness responses of the 3D printed samples. For the layer resolution, 0.3 mm layer height is associated with the highest mean roughness response and 0.1 mm with the lowest roughness mean response.

Analysis of variance (ANOVA) of surface Rockwell microhardness

A similar analysis was undertaken for surface hardness values and the results are shown in Figs. 2.8 and 2.9. According to the mean plots, the responses for both factors are not the same at all levels indicating the presence of the main effect. At 45°

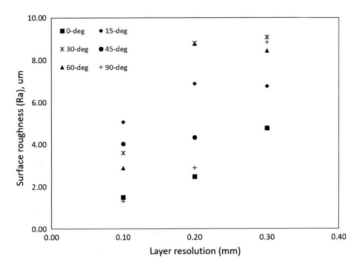

Fig. 2.5 Variation of the average surface roughness with the layer resolution

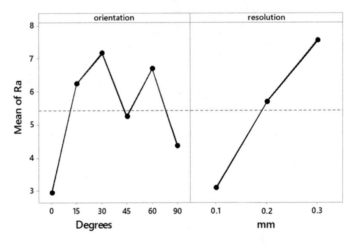

Fig. 2.6 Main effect plots of surface roughness against the resolution and orientation angles

orientation and 0.3 mm resolution, the highest mean Rockwell hardness responses are obtained whereas the lowest means of hardness are obtained at 90°orientation and 0.1 mm layer resolution. Analysis of variance (ANOVA), however, indicated that both factors are statistically insignificant (all p-values are larger than 0.05) to surface hardness properties of the 3D printed PLA samples. It is can, however, be deduced that the build orientation of the print is closer to significance than the layer resolution. The summary of the model clearly shows a larger error to regression fitting as compared to the roughness ANOVA model in Table 2.4.

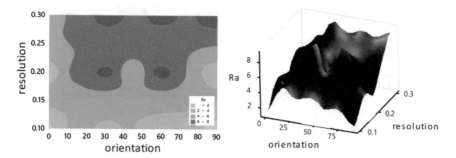

Fig. 2.7 Surface plots showing the relationship among surface roughness (Ra), layer resolution and build orientation

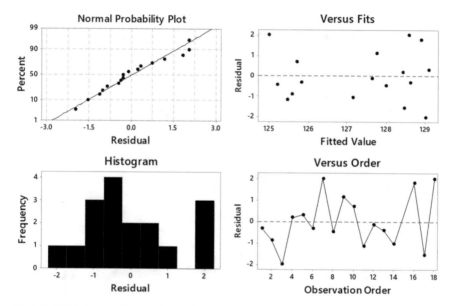

Fig. 2.8 Validation tests for experimental assumptions for average Rockwell microhardness values

Interaction analysis

Furthermore, an interaction analysis of the two factors at their respective levels was undertaken and the results are presented as interaction plots in Figs. 2.10 and 2.11 for mean surface roughness and mean hardness responses, respectively. In both cases, the interaction plots are non-parallel indicating that there is a relationship between the two factors. It is important to note that the interactions are not very strong since some of the line graphs are parallel to each other. However, it can be deduced that the lowest roughness of the PLA printing can be obtained at a layer resolution of 0.1 mm and build orientations of 0° and 90°. On the contrary, the highest roughness can be obtained with a resolution of 0.3 mm at build orientations of 30° and 90°.

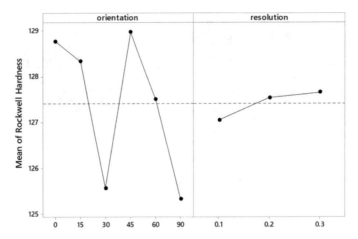

Fig. 2.9 Main effect plots of Rockwell hardness values against the angle of orientation (°) and resolution (mm)

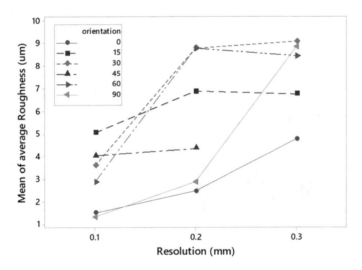

Fig. 2.10 Mean interaction plots of resolution at various build orientations angles for roughness response

Considering the mean hardness response, the highest mean hardness can be obtained at a layer resolution of 0.2 mm and a build orientation of 0° whereas the lowest hardness can be obtained at a layer resolution of 0.2 mm and an orientation of 90°. The results are further presented in surface plots shown in Fig. 2.12. A ranking of the microhardness values of the full factorial experimental design is further presented in Tables 2.7 and 2.8.

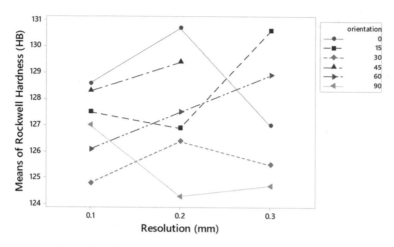

Fig. 2.11 Mean interaction plots of resolution at various build orientations angles for Rockwell hardness response

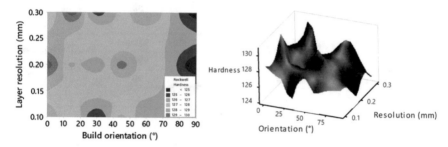

Fig. 2.12 Surface plots showing the relationship among Rockwell hardness, layer resolution and build orientation

Table 2.7 One-way analysis of variance (ANOVA) results for surface roughness response

Source	DF	Adj SS	Adj MS	F-value	P-value
Model	7	36.33	5.19	1.97	0.169
Linear	7	36.33	5.19	1.97	0.169
Orientation	5	35.62	7.12	2.71	0.092
resolution	2	1.13	0.57	0.22	0.810
Error	9	23.70	2.63		
Total	16	60.03			
	Model summary				
	Standard deviation	R^2	R^2(adj)	R^2(pred)	
	1.62	60.5%	29.82%	0.00%	

Table 2.8 Ranking of the Rockwell hardness values according to the interaction between the 3D printing parameters, i.e. build orientation and layer resolution

Rank	Layer resolution (mm)	Build orientation (°)
Highest mean hardness	0.2	0
	0.3	15
	0.2	45
	0.1	0
	0.1	45
	0.3	60
	0.2	60
	0.1	15
	0.1	90
	0.3	0
	0.2	15
	0.2	30
	0.3	30
	0.1	60
	0.1	30
	0.3	90
Lowest mean hardness	0.2	90

2.4   Summary

In this chapter, the strategy of enhancing the quality of 3D printed PLA samples by optimizing layer resolution and build orientation is presented. A mixed full factorial design of experiment ($6^1 \times 3^1$) of the two printing parameters at one replicate was undertaken. Eighteen PLA samples with dimensions of 10 mm by 15 mm by 5 mm were printed using a desktop (FDM) 3D printer. The layer resolution levels used were 0.1 mm, 0.2 mm and 0.3 mm whereas the levels of build orientation used were 0°, 15°, 30°, 45°, 60° and 90°. The other parameters such as temperature, infill density and speed were kept constant throughout the FDM process. The samples were then characterized for surface roughness and microhardness using handheld hardness tester and Rockwell microhardness, respectively. These properties were used as the responses for the full factorial analysis of the results. The main effects plots and analysis of variance (ANOVA) revealed that the layer resolution is the most significant factor in determining the surface roughness of the PLA printed samples. It was also shown that the two factors are not statistically significant on the Rockwell surface microhardness although build orientation is closer to influencing the hardness of the samples as compared to the resolution. The mean interaction plots revealed that a resolution height of 0.1 mm and build orientations of 0° and 90° produce samples with the lowest mean roughness. The samples prepared at a layer resolution and orientation of 0.2 mm and 0°, respectively, exhibited the highest surface hardness characteristics. These results also showed that there is no strong statistical interaction between layer resolution and build orientation, however, 0° and 90° are the extreme orientations which influence the quality of the 3D prints at various layer resolutions.

References

1. J.S. Shim, J.E. Kim, S.H. Jeong, Y.J. Choi, J.J. Ryu, Printing accuracy, mechanical properties, surface characteristics, and microbial adhesion of 3D-printed resins with various printing orientations. J. Prosthet. Dent. 1–8 (2019)
2. T. Yao, J. Ye, Z. Deng, K. Zhang, Y. Ma, H. Ouyang, Tensile failure strength and separation angle of FDM 3D printing PLA material: Experimental and theoretical analyses. Compos. Part B Eng. 107894 (2020)
3. Y.T. Wang, Y.T. Yeh, Effect of print angle on mechanical properties of FDM 3D structures printed with POM material. In: *Lecture Notes in Mechanical Engineering*, vol. PartF9 (2017), pp. 157–167
4. N.H. Patadiya, H.K. Dave, S.R. Rajpurohit, Effect of build orientation on mechanical strength of FDM Printed PLA. In: *Proceedings of AIMTDR* (2020), pp. 301–307
5. M.F. Afrose, S.H. Masood, P. Iovenitti, M. Nikzad, I. Sbarski, Effects of part build orientations on fatigue behaviour of FDM-processed PLA material. Prog. Addit. Manuf. 1(1–2), 21–28 (2016)
6. V. Kovan, G. Altan, E.S. Topal, Effect of layer thickness and print orientation on strength of 3D printed and adhesively bonded single lap joints. J. Mech. Sci. Technol. 31(5), 2197–2201 (2017)
7. M. Samykano, S.K. Selvamani, K. Kadirgama, W.K. Ngui, G. Kanagaraj, K. Sudhakar, Mechanical property of FDM printed ABS: influence of printing parameters. Int. J. Adv. Manuf. Technol. 102(9–12), 2779–2796 (2019)
8. S. Deswal, R. Narang, D. Chhabra, Modeling and parametric optimization of FDM 3D printing process using hybrid techniques for enhancing dimensional preciseness. Int. J. Interact. Des. Manuf. 13(3), 1197–1214 (2019)
9. K. Logesh, V.K.B. Raja, M. Venkatasudhahar, H.K. Rana, *Innovative Design, Analysis and Development Practices in Aerospace and Automotive Engineering (I-DAD 2018)* (Springer, Singapore, 2019)
10. M.S. Saad, A.M. Nor, M.E. Baharudin, M.Z. Zakaria, A.F. Aiman, Optimization of surface roughness in FDM 3D printer using response surface methodology, particle swarm optimization, and symbiotic organism search algorithms. Int. J. Adv. Manuf. Technol. 105(12), 5121–5137 (2019)
11. O.A. Mohamed, S.H. Masood, J.L. Bhowmik, M. Nikzad, J. Azadmanjiri, Effect of process parameters on dynamic mechanical performance of FDM PC/ABS printed parts through design of experiment. J. Mater. Eng. Perform. 25(7), 2922–2935 (2016)
12. F.M. Mwema, E.T. Akinlabi, O.S. Fatoba, Visual assessment of 3D printed elements : A practical quality assessment for home-made FDM products. Mater. Today Proc. (2020)
13. T. Kivak, Optimization of surface roughness and flank wear using the Taguchi method in milling of Hadfield steel with PVD and CVD coated inserts. Meas. J. Int. Meas. Confed. 50(1), 19–28 (2014)
14. F.M. Mwema, E.T. Akinlabi, O.P. Oladijo, Fractal analysis of hillocks: A case of RF sputtered aluminum thin films. Appl. Surf. Sci. 489, 614–623 (2019)

15. F.M. Mwema, E.T. Akinlabi, O.P. Oladijo, Effect of substrate type on the fractal characteristics of AFM images of sputtered aluminium thin films. Mater. Sci. **26**(1), 49–57 (2019)
16. F.M. Mwema, E.T. Akinlabi, O.P. Oladijo, S. Krishna, J.D. Majumdar, Microstructure and mechanical properties of sputtered Aluminum thin films. Procedia Manuf. **35**, 929–934 (2019)
17. F.M. Mwema, E.T. Akinlabi, O.P. Oladijo, Complementary investigation of SEM and AFM on the morphology of sputtered aluminum thin films. In: *Proc. of the Eighth Intl. Conf. on Advances in Civil, Structural and Mechanical Engineering—CSM 2019* (2019), pp. 10–14
18. Ş. Ţălu, I.A. Morozov, R.P. Yadav, Multifractal analysis of sputtered indium tin oxide thin film surfaces. Appl. Surf. Sci. **65**(3), 294–300 (2019)
19. M.F. Erinosho, E.T. Akinlabi, O.T. Johnson, Characterization of surface roughness of laser deposited titanium alloy and copper using AFM. Appl. Surf. Sci. **435**, 393–397 (2018)

Chapter 3
Multi-objective Optimization Strategies

Abstract In this chapter, multi-objective optimization as a strategy for quality production of parts through fused deposition modelling is presented. Various techniques used in undertaking the multi-objective optimization process are described based on case studies from the literature and the authors' data. The general algorithms for multi-objective optimization of the FDM process are described. The most significant objectives of the various optimization cases are identified and described in relation to the quality of the fused deposition modelling of parts. The main objectives for optimizing fused deposition process are (i) to increase the rate of production, (ii) to reduce material wastage and utilize as minimum material as possible, (iii) save on the cost of power consumption during printing and (iv) achieve the highest quality of FDM parts.

Keywords 3D printing · Fused deposition modelling · Genetic algorithms · Grey relational degree · Multi-objective optimization · Pareto · Printing parameters

3.1 Introduction

Optimization generally involves determining the maximum or minimum value considering one or several objectives. In cases where several objectives are involved, the problem is known as a multi-objective optimization (MOO). The objectives of any project or process are generated based on the problems or limitations associated with it. For instance, in a typical construction project, the challenges involved are budget and time constraints, safety, health, quality, etc. As such, a construction project involves multiple objectives including productivity maximization, safety and health, minimum duration and cost. The combination of these objectives should be considered for an optimal solution to the construction project. The approach of MOO is adopted over a single objective since in most real industrial processes, optimizing one aspect (parameter) has a direct influence on the other parameters and therefore could cause conflicts among the optimized and other parameters [1].

In the fused deposition modelling (FDM) process, the objective is to achieve low surface roughness, high mechanical strength, low defect density (such as cracks

© The Author(s), under exclusive license to Springer Nature Switzerland AG 2020
F. M. Mwema and E. T. Akinlabi, *Fused Deposition Modeling*,
Manufacturing and Surface Engineering,
https://doi.org/10.1007/978-3-030-48259-6_3

and porosity), thermally stable prints, dimensionally accurate products, reduce on printing time, material cost, just to mention a few. As such, just like other manufacturing processes, fused deposition modelling is a multi-objective problem and requires the MOO approach for optimal process and product quality. There are so many MOO methods utilized across various fields, some of them include Genetic algorithms (GA), Differential evolution (DE), Pareto evolutionary algorithm (PEA), Non-dominated sorting genetic algorithm-II (NSGA-II), Particle swarm optimization (PSO), Hungarian algorithm (HA), Analytic network process (ANP), Hybrid methods, among so many other techniques.

3.2 Theory of Multi-optimization Techniques

Multi-objective optimization (MOO) deals with minimization or maximization of a vector objective $F(x)$ subject to equality ($h_l(x)$) and inequality $g_k(x)$) constraint functions. The problem for MOO is generally formulated as illustrated by Eqs. (3.1) and (3.2).
  Optimize

$$f(x) = [f_1(x), f_2(x), \ldots f_k(x)]^T \tag{3.1}$$

$$f(x) \in R^k, (x) \in R^n \tag{3.2}$$

Subject to equality and inequality constraint functions as defined by Eqs. (3.3) and (3.4).

$$g_k(x) \leq 0, k = 1, 2, \ldots m \tag{3.3}$$

$$h_l(x) \leq 0, l = 1, 2, \ldots p \tag{3.4}$$

The $F(x)$ is also subject to side constraints (x_j^{inf} and x_j^{sup}) in n-dimensional real space defined in Eq. (3.5).

$$x_j^{\text{inf}} \leq x_j \leq x_j^{\text{sup}}; j = 1, 2, \ldots n \tag{3.5}$$

where x is a vector of decision variables $[x_1, x_2, \ldots, x_n]$ and the problem is considered in two-dimensional Euclidean space (where n is the dimensional space of the decision variables and k is the dimensional space of the objective functions). The objective is to determine the set of vectors which satisfy g_k and h_l and the particular set of values of $x_1, x_2, \ldots, x_n$ which yield the optimum values of all the objective functions. These constraints define the feasible region as shown by Eq. (3.6).

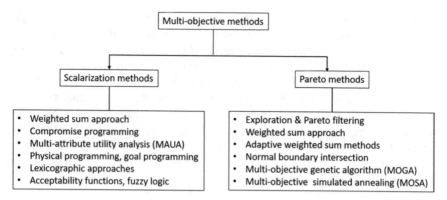

Fig. 3.1 Classification of multi-objective optimization (MOO) methods

$$\Omega = \left\{x \in R^n / g_{(x)} \leq 0; \; h_{(x)} = 0\right\} \tag{3.6}$$

In most cases, the components of the vector $F(x)$ (objective function) will compete with each other and therefore, such problems will have several solutions. The dilemma lies in the choice of the solution for the optimization problem. To determine an optimal solution, one must find the minimum attainable for all the objective functions separately. The Pareto method of non-dominated solutions is one of the methods to determine the optimal solution for a set of competing objective functions [2]. Pareto optimality enables us to determine the 'trade-offs' rather than single solutions for multi-objective problems [3]. Some of the classical methods of multi-objective optimization include scalarization (weighting), hierarchical, trade-off, global criterion and goal programming methods [3]. Multi-objective evolutionary algorithms (MOEAs) have been developed to advance these classical methods. According to Weck (2004), the MOO techniques can be broadly classified into scalarization and Pareto methods as summarized in Fig. 3.1 [4]. The two methods are briefly described here to provide insights into the basics of MOO process and this is illustrated in the next section with case studies of fused deposition modelling.

3.2.1 Pareto Methods

The Pareto method determines the most efficient solutions from a set of feasible solutions which arises from conflicting objective functions. Pareto improvement involves movement of one feasible solution to another that can make (i) at least one objective function to return a better value and (ii) with no other objective function becoming worse off. In this process, the elements of the solution vectors are kept independent from each other and the principle of dominance is adopted to differentiate the dominated and non-dominated solutions. Once the solutions are chosen such that changes in one objective function influence the other objective function, the solution is called

Pareto optimal solution. Generally, Pareto optimality solutions are obtained as per the following major steps: (i) Determination of the feasible solutions for maximization or minimization of objective functions, (ii) Undertaking non-dominating sorting, (iii) Determination of Pareto front and assignment of ranks to populations of solutions and finally (iv) For each rank of population, determine the crowding distance. Usually, solutions in rank 1 are non-dominated and dominate all the other sets in the solution populations. Additionally, solutions in each rank have the same fitness and solutions in rank 1 exhibit the highest fitness. In determining the 'preferred' solution from each rank, sets lying in less-crowded areas are chosen and this means that those with the largest crowding distance are chosen. Readers are referred to [5] for further description of the Pareto method.

3.2.2   Scalarization Method

In this method, the multi-objective problem is converted into a single-objective solution before the optimization process starts. The objective function is assigned various contributions (weights) to form a weighted sum of all the objectives as shown in Eqs. (3.7) and (3.8).

$$f(x) = \sum_{i=1}^{k} w_i f_i(x) r_i \tag{3.7}$$

where

$$\sum_{i=1}^{k} w_i = 1 \tag{3.8}$$

r_i are constant multipliers, $w_i \geq 0$ are weighting coefficients that show the relative significance of each choice. Usually, the challenge is attaching the weights to various objectives and determining their importance. It is important to ensure that the units of the weights are approximately the same as the numerical values of all the functions. The best results are obtained if the multiplier (r_i) is an inverse of the ideal solution (f_i^0).

3.3   Case Studies in Optimization of FDM

As mentioned, fused deposition modelling (FDM) can be approached as a multi-objective optimization problem. For example, just like any other manufacturing process, the objective of any FDM manufacturer is to minimize production time, cost and material wastage. In this section, some case studies (from the literature) on multi-objective optimization of the FDM processes are discussed with emphasis on the MOO techniques utilized. The overall aim is to illustrate the MOO approach as a strategy for quality enhancement in FDM manufacturing.

3.3.1 Case 1: Non-dominated Sorting Genetic Algorithm-II (NSGA-II)

A study by Asadollahi-Yazdi et al. [6], titled 'Multi-Objective Optimization of Additive Manufacturing Process', is one of the classic case studies in the multi-objective optimization of fused deposition modelling (FDM). There were two objective functions formulated in the study namely, production time and material mass, two decision variables namely layer thickness and part orientation were chosen, and surface roughness and mechanical strength of the prints were considered as the constraint functions. The objective of any manufacturing process, including FDM, is to minimize the time of production and material utilization to achieve the lowest cost. These two objectives are influenced by process parameters of which the literature has shown that orientation and layer thickness are the most significant and as such were chosen as the decision variables in this research. However, the extent to minimize both time and material utilization during the FDM process is constrained by surface roughness (Ra) and mechanical strength (UTS) of the printed product. The orientation angles $(\theta_x, \theta_y$ and $\theta_z)$ of the part in space in relation to X-, Y- and Z-axes were specified as follows: varying between -180° and 180° for X- and Z-axes while for Y-axis was chosen between 0° and 180°. With the minimum and maximum layer thicknesses defined as L_{tmin} and L_{tmax}, respectively, the multi-objective optimization problem of this study was formulated according to Eqs. (3.9)–(3.15):

Minimize

$$f_1(x) = \text{Time}(x) \tag{3.9}$$

$$f_2(x) = \text{Material}(x) \tag{3.10}$$

Under the constraints:

$$g R_a(x) \leq R_{amax} \cdots \text{(surface roughness constraint)} \tag{3.11}$$

$g_{UTS}(x) \geq \sigma_{max}$ (mechanical behaviour of AM products, ultimate tensile strength (UTS)).

$$l_b \leq x \leq u_b \tag{3.12}$$

With

$$x = \left[\theta_x, \theta_y, \theta_z, L_t\right] \tag{3.13}$$

$$l_b = \left[-180°, 0°, -180°, L_{\text{tmin}}\right] \tag{3.14}$$

$$u_b = \left[180°, 180°, 180°, L_{\text{tmax}}\right] \tag{3.15}$$

The upper and lower bounds of the decision variable (x) were defined as u_b and l_b, respectively. Based on the various parameters, an experimental study involving slicing and 3D printing of the parts at different orientations and layer thicknesses was undertaken. The manufacturing time and material usage were determined from the slicing software whereas the constraints were measured on the printed parts, i.e. surface roughness and mechanical strength. Non-dominated sorting genetic algorithm II (NSGA-II) technique was applied to determine the optimal manufacturing conditions of the sample printed product (bag hook in this case). The generated population of solutions was evaluated through the Pareto optimal non-dominated front plots for the two objective functions followed by crowding distance computations for different rankings of solutions. The approach was shown to be effective for achieving optimal manufacturing cost and quality products during the FDM process.

3.3.2   Case 2: Signal-to-Noise and Grey Correlation Degree Multi-objective Optimization

A 'multi-objective optimization of process parameters for biological 3D printing composite forming based on SNR and grey correlation degree' undertaken by Jiang et al. in 2015 [7] is another important case study. The study aimed at achieving quality objectives to enhance the quality of 3D printed scaffolds, namely, wire width and layer height errors. The study utilized signal-to-noise ratio to compute the uncertainties in the process and grey relational method for multi-objective optimization. The study was based on an orthogonal multilevel method consisting of six parameters and five levels ($L_{25}(5^6)$ orthogonal array). The parameters considered were platform velocity, extrusion speed, fibre spacing, electrospinning concentration, acceptance distance and voltage. The multi-objective optimization problem in this study was formulated as follows.

The objective of the optimization was to minimize the errors in width and layer height during the 3D printing of biological scaffolds. As such, the formulation is represented by Eqs. (3.16) and (3.17).

Minimize

$$f_1(x) = w_{\text{error}}(x_i); \ i = 1, 2, \ldots, 25 \tag{3.16}$$

$$f_2(x) = l_{error}(x_i); \, i = 1, 2, \ldots, 25 \tag{3.17}$$

where $f_1(x)$ and $f_2(x)$ denote the objective functions for width error (w_{error}) and layer height error (l_{error}), respectively. From the objective functions, x_i indicates the actual values of the objective functions at each experiment and i denotes the number of the test runs, which varied between 1 and 25. A set of measurements were obtained on the 25 experiments according to the design of the experiment (orthogonal array) from which the errors in width and layer height were determined. The signal-to-noise ratios for each of the objective functions at each experiment were determined using the 'smaller-the-best' criterion since the aim was to minimize the errors.

The grey relation modelling was undertaken on the SNR results for the two objective functions as illustrated in the chart in Fig. 3.2.

As shown in Fig. 3.2, the grey relation coefficient (GRC), grey relation grade (GRG) and average grey correlation degree were the outputs of the grey relation model. Based on the model and SNR methodology, the following conclusions can be deduced, which can be adopted as strategies for quality enhancement of the FDM process.

- The sequence of the influence of the specific control parameters was determined from which the material extrusion speed was identified as the most influential parameter determining the accuracy of the 3D printing of biological scaffolds. The sequence of the other parameters was as follows: platform velocity, accepting distance, electrospinning concentration, voltage and fibre spacing. The voltage and fibre spacing were the least significant parameters influencing the accuracy of the scaffolds. A survey of the literature shows very few studies investigating the influence of these two parameters on the quality of FDM products. This could be attributed to the fact that the parameters have minimal influence on the process and hence the quality of the products. The speeds of both the extruder and the build plate are identified as very important factors during FDM. These parameters affect melting and fusion of the filament during the process. If proper speeds are not chosen, the staircasing effect and formation of cusps may occur resulting in improper fusion and hence inaccurate height and width of the layers. In fact, the

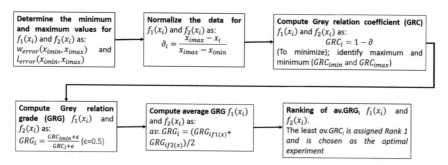

Fig. 3.2 Summary of the grey relation model optimization adopted by Jiang et al. [7]

layer thickness has a direct relationship with the staircase; that is, the larger the layer thickness, the larger the staircase effect and hence, the poor the surface quality of the prints [8].

- According to the grey relation model, the optimum combination of the parameters was extrusion speed of 16 mm/s, plate speed of 16 mm/s, electrospinning concentration of 8.7%, voltage of 19 kV, fibre spacing and accepting distance of 1.4 mm and 100 mm, respectively. The model optimization was confirmed through an experiment undertaken under the optimum conditions and high-quality biological scaffolds were produced.

3.3.3   Case 3: Particle Swarm Optimization Method for Fused Deposition Modelling Process

A particle swarm optimization (PSO) method was utilized by Dey, Hoffman and Yodo in 2019 in their paper titled, 'optimizing multiple process parameters in FDM with PSO' [9]. In their investigation, four decision variables namely build orientation, infill density, extrusion temperature and layer thickness were tested. Additionally, two objective functions namely compressive strength and printing time were used. As mentioned earlier, the objective FDM process is to reduce the manufacturing time and enhance the strength of the printed products. These are two major aspects limiting industrial and mass production adoption of additive manufacturing technologies. Each of the process parameters had three levels; a face-centred central composite design (FCCCD) design of the experiment was applied to the array (3^4). This approach belongs to a group of DOE methods called response surface model (RSM). As such, the minimum and optimal number of experiments used from the FCCCD design was 30 experiments.

Particle swarm optimization (PSO) is a meta-heuristic optimization procedure that is inspired by natural behaviour of birds, fish and plants in their natural setting especially in search of water, food and sunlight and it was developed by Kennedy and Eberhart in 1995 [10]. The algorithm takes each of the solution candidates in the solution space as a particle in a swarm and the optimization involves the iterative improvement of each candidate with respect to a given measure of quality. Within the solution space, each particle has velocity and position which the algorithm uses to obtain the optimal points. The equations describing position (X) and velocity (V) are represented by Eqs. (3.18) and (3.19), respectively.

$$V_i(t + 1) = \omega V_i(t) + c_i r_i(t)(P_i(t) - X_i(t)) + c_2 r_2(t)(g(t) - X_i(t)) \qquad (3.18)$$

$$X_i(t + 1) = X_i(t) + V_i(t + 1) \qquad (3.19)$$

where ω is the inertia coefficient, c_i and c_2 represent the cognitive and social acceleration coefficients (learning factors) respectively, r_1 and r_2 are random numbers

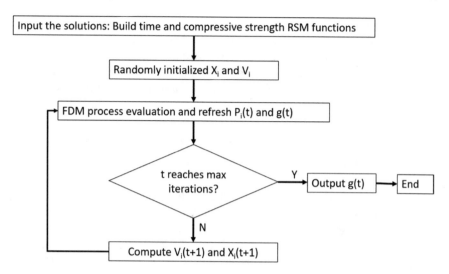

Fig. 3.3 Flow chart of the PSO methodology utilized by [9]

varying between 0 and 1. Also, $P_i(t)$ and $g(t)$ represent the best location (best experience) of each particle and the best common location (global best) for all the particles, respectively. In fact, the function $g(t)$ represents the minimum global (optimal) point of all the particles in the swarm.

In this study, the following step-by-step procedure (Fig. 3.3) for PSO multi-objective optimization was implemented according to study [11].

As shown in Fig. 3.3, the response surface quadratic modelling was applied to determine the relationship among the FDM parameters and the compressive strength and build time. It was observed that the extrusion temperature was insignificant for both compressive strength and printing time. As such, the extrusion temperature was eliminated among the decision variables of the multi-objective problem. The quadratic equations (relating the decision variables to the build time and compressive strengths) were developed, which were then used as the objective functions of the particle swarm optimization. In the optimization problem, the layer thickness (mm), orientation (°) and density (%) were the decision variables whereas the parameter levels (low, average and high) were taken as the constraints. The algorithm was implemented based on the pseudocodes reported in studies [12, 13]. The result of the multi-objective particle swarm optimization (MOPSO) was a Pareto frontier to represent the trade-off between the conflicting objectives. Two main conclusions can be deduced from this study for enhanced strength and optimal time of printing of PLA samples:

- For the set of non-dominating solutions obtained, the build orientation of 0° was the best for both objectives. At this orientation, the fibres of the PLA are layered parallel to the direction of the compressive load and therefore they provide the best resistance to loading as compared to 45° and 90°.

- At $0°$ printing orientation, the printer achieves the required thickness at a lower number of layers. This is because the build plate is only required to move in a downward direction when printing the subsequent layers. In other orientations, the plate may be required to incline or turn besides the downward movement, which requires a longer time for the printer to form the subsequent layers.

3.3.4 Case 4: Full Factorial and Grey Relational Degree Optimization of FDM Printed PLA

In this section, we present our multi-objective optimization of the study presented in Chap. 2 of this book. The experiment for the PLA printing was undertaken using a full factorial mixed-level design. The layer thickness (three levels) and build orientation (six levels) are the decision variables whereas the levels are the constraints. The objective of the process is to minimize the roughness (Ra) and printing time (t) and maximize the surface hardness (HBR). Therefore, the multi-objective optimization problem in this case is formulated according to Eqs. (3.20) and (3.21).

Objective functions: Minimize

$$f_1(x) = Ra_i(x) \tag{3.20}$$

$$f_2(x) = t_i(x) \tag{3.21}$$

Maximize

$$f_3(x) = HBR_i(x); \text{ for all the objective functions, } i = 1, 2, \ldots .18$$

The functions are subject to the following constraints of different levels of each function.

For layer thickness, $0.1 \leq x \leq 0.3$

For build orientation , $0° \leq x \leq 90°$

As detailed in Chap. 2, 18 experiments were undertaken at given constraints. However, our printer did not print at one of the conditions ($45°$ build orientation and 0.3-mm layer resolution) and therefore there were only 17 investigations undertaken. The time of printing was determined directly from the slicing software whereas the roughness and hardness tests were undertaken on respective laboratory facilities. The obtained values of the decision variables for each function index (i) are shown in Table 3.1. A grey relational modelling was then applied to this data following the

Table 3.1 The experimental order, decision variables and objective functions

Chapter 2 exp. number	Resolution	Orientation	Ra	HBR	t (min)
1	0.1	30	3.61	124.8	24
2	0.2	15	6.88	126.9	13
3	0.1	60	2.89	126.1	24
4	0.1	45	4.04	128.3	24
5	0.1	15	5.07	127.5	24
6	0.2	30	8.8	126.4	14
7	0.2	60	8.77	127.5	13
8	0.2	45	4.35	129.4	10
10	0.1	90	1.33	127	25
11	0.2	90	2.89	124.3	15
12	0.3	60	8.44	128.9	7
13	0.3	0	4.78	127	15
14	0.2	0	2.48	130.7	18
15	0.1	0	1.51	128.6	28
16	0.3	15	6.76	130.6	10
17	0.3	30	9.08	125.5	7
18	0.3	90	8.84	124.7	11
Minimum (X_{min})			1.33	124.3	7
Maximum (X_{max})			9.08	130.7	28

general methodology in Fig. 3.2. In Table 3.1, the minimum and maximum values for each objective function were identified and are stated in the last rows of the table.

Using the minimization equation in Fig. 3.2, the normalization was undertaken for Ra and t functions whereas the normalization for maximization of HBR function was undertaken using Eq. (3.22).

$$\text{Normalization for maximization of HBR} = \frac{X_i - X_{min}}{X_{max} - X_{min}} \qquad (3.22)$$

Next, the grey relation coefficient (GRC) was determined for the three functions as earlier described in Fig. 3.2. However, it should be noted that for the maximization process, the values of the normalized function (HBR) represent the GRC and therefore, no transformation was applied in the data. The results of these operations are shown in Table 3.2. It can be noted that the column of HBR data is the same for both sets of data (normalized and GRC data) since it is a maximization objective.

Next, the data in Table 3.2 was transformed into a grey relational grade (GRG) and degree (rank) following the equations in Fig. 3.2. The value of $\in = 0.5$ was used in this experiment and its choice was based on the existing literature and the results of the operations are shown in Table 3.3. As shown, the grey relation model transformed

Table 3.2 Normalized data for objective functions and computation of grey relational coefficient

Normalized data				Grey relational coefficient (GRC)		
Exp. number	Ra	HBR	t (min)	Ra	HBR	t (min)
1	0.7058	0.0781	0.1905	0.2942	0.0781	0.8095
2	0.2839	0.4063	0.7143	0.7161	0.4063	0.2857
3	0.7987	0.2813	0.1905	0.2013	0.2813	0.8095
4	0.6503	0.6250	0.1905	0.3497	0.6250	0.8095
5	0.5174	0.5000	0.1905	0.4826	0.5000	0.8095
6	0.0361	0.3281	0.6667	0.9639	0.3281	0.3333
7	0.0400	0.5000	0.7143	0.9600	0.5000	0.2857
8	0.6103	0.7969	0.8571	0.3897	0.7969	0.1429
10	1.0000	0.4219	0.1429	0.0000	0.4219	0.8571
11	0.7987	0.0000	0.6190	0.2013	0.0000	0.3810
12	0.0826	0.7188	1.0000	0.9174	0.7188	0.0000
13	0.5548	0.4219	0.6190	0.4452	0.4219	0.3810
14	0.8516	1.0000	0.4762	0.1484	1.0000	0.5238
15	0.9768	0.6719	0.0000	0.0232	0.6719	1.0000
16	0.2994	0.9844	0.8571	0.7006	0.9844	0.1429
17	0.0000	0.1875	1.0000	1.0000	0.1875	0.0000
18	0.0310	0.0625	0.8095	0.9690	0.0625	0.1905
Minimum				0	0	0
Maximum				1	1	1

the three objectives into a single-objective optimization problem (average GRG). Then, the ranking (degree) shows the sequence of the significance of each index (i) in the optimization. The highest degree (rank 1) shows the most optimal setpoint of the experiment and the best trade-off.

From Table 3.3, it can be seen that the most optimal set of the solution was obtained at index ($i = 11$) or experiment no. 11 in which the values of roughness (Ra), hardness (HBR) and time of manufacturing (t) were 2.89 μm, 124.3, and 15 min, respectively (see Table 3.1). The results in Table 3.3 represent the Pareto optimality and non-dominating solutions for a multi-objective optimization problem. The result represented by rank 1 is, therefore, the Pareto optimal (non-dominated front), followed by rank 2 and so forth.

To demonstrate the relationship between the results in Table 3.3 and Pareto optimal and non-dominating frontier concepts, a three-dimensional scatter plotting of the values of roughness, hardness and time from Table 3.1 was undertaken. In this case, it was assumed that the actual laboratory values of the three objective functions represent the solution space for the multi-objective problem. Figure 3.4 shows the 3D scatter plots of the three responses. As shown in Fig. 3.4 and Table 3.3, the non-dominated frontier provides the best trade-offs among the objectives of the

Table 3.3 Computation of grey relational grade (GRG), average GRG and degree (rank)

Exp. number	Resolution	Orientation	Grey relational grade (GRG)			Average GRG	Rank
			Ra	HBR	t (min)		
1	0.1	30	0.6296	0.8649	0.3818	0.6254	5
2	0.2	15	0.4111	0.5517	0.6364	0.5331	11
3	0.1	60	0.7130	0.6400	0.3818	0.5783	7
4	0.1	45	0.5885	0.4444	0.3818	0.4716	16
5	0.1	15	0.5089	0.5000	0.3818	0.4636	17
6	0.2	30	0.3416	0.6038	0.6000	0.5151	13
7	0.2	60	0.3425	0.5000	0.6364	0.4929	15
8	0.2	45	0.5620	0.3855	0.7778	0.5751	8
10	0.1	90	1.0000	0.5424	0.3684	0.6369	4
11	0.2	90	0.7130	1.0000	0.5676	0.7602	1
12	0.3	60	0.3528	0.4103	1.0000	0.5877	6
13	0.3	0	0.5290	0.5424	0.5676	0.5463	10
14	0.2	0	0.7711	0.3333	0.4884	0.5309	12
15	0.1	0	0.9556	0.4267	0.3333	0.5719	9
16	0.3	15	0.4164	0.3368	0.7778	0.5104	14
17	0.3	30	0.3333	0.7273	1.0000	0.6869	2
18	0.3	90	0.3404	0.8889	0.7241	0.6511	3

Fig. 3.4 Graphical representation of the solution candidates for the multi-objective optimization problem. The Pareto optimality (non-dominated fronts) are also shown as ranks 1, 2 and 3

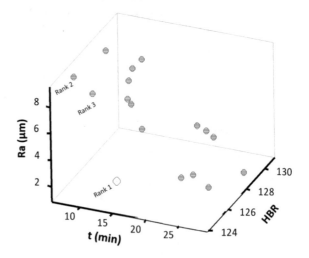

optimization process. For example, comparing ranks 1, 2 and 3, it can be seen that experiment 11 dominates rank 2 (experiment 17) in terms of low roughness and high hardness except for the time of production. It can also be seen that rank 1 (experiment 11) dominates rank 3 (experiment 18) in all the objectives, i.e. using experiment 11 conditions produces FDM components at low roughness, high hardness and at a higher rate of production as compared to experiment 18. Therefore, according to the grey relation and Pareto non-dominated front models, 3D printing of PLA at a build orientation of 90° and layer height (resolution) of 0.2 mm provides the highest fitness and best conditions for low roughness, high production rate (time) and better mechanical strength (hardness).

In a related approach, grey relational models can be used in conjunction with the Taguchi optimization method to determine the influence and significance of the two decision variables used during the fused deposition modelling. The results of the GRG (reported in Table 3.3) were taken as the responses to the Taguchi optimization model and the following results were obtained (Fig. 3.5 and Table 3.4). The optimization was based on the 'larger-the-best'. As shown, the highest S/N ratios were observed at 90° and 0.3-mm build orientation and layer height (resolution), respectively. These parameters coincide with experiment 18, which was ranked third as per the grey relation modelling. The results of the ANOVA (at 95% confidence level) for the GRG show that the resolution has an insignificant influence on the GRG and hence to the roughness, time and hardness as compared to the build orientation. The P-values for orientation were around 0.03 whereas that of the resolution was 0.65. The plots of means of GRG for the two parameters are also shown in Fig. 3.6 and it further affirms the Taguchi model in this study.

The most important aspect of the Taguchi is that it undertakes a single-objective optimization and it is able to rank the significance of each processing parameter's

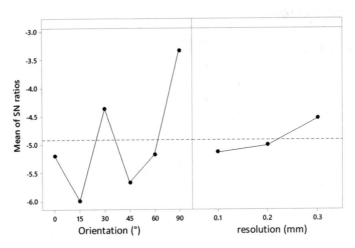

Fig. 3.5 Main effects plots for S/N ratios of the grey relation grade (GRG) for the 3D printing of PLA samples at different orientation angles and layer resolution

Table 3.4 S/N response table for GRG. The bold values indicate the maximum S/N ratios of the Taguchi optimization

Level	Orientation	Resolution
1	−5.201	−5.136
2	−5.995	−5.009
3	−4.367	−4.540
4	−5.667	
5	−5.173	
6	−3.342	
Delta	2.653	0.595
Rank	1	2

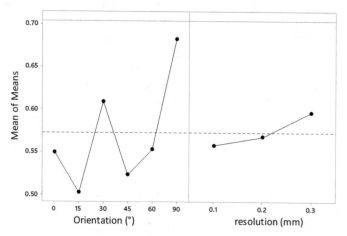

Fig. 3.6 The main effect plots of means of GRG for the 3D printed PLA samples at different levels of orientation and layer resolution

level to the quality of the 3D printing. Here, the most influential level for orientation was 90° followed by 30°, 0° and 60°, while the least significant levels were 45° and 15°. For the layer thickness, the most influential level was 0.3 while the least important level was 0.1 mm. Although the products printed at higher layer resolution exhibited very high roughness, they were shown to have better hardness and lower time of printing. At a larger layer thickness, higher rate of filament material is extruded and hence the required print thickness is achieved quickly as compared to when a smaller layer resolution is utilized. Furthermore, during the manufacturing of rectangular samples, such as those illustrated in this work, either 0° or 90° angles are observed as the best printing orientations. However, their choice will depend on layer resolution and the best trade-off for the decision makers. A trade-off has to be made between productivity rate and surface quality/strength of the samples (Table 3.5).

Table 3.5 Results of ANOVA for grey relation grade (GRG)

Source	DF	Adj SS	Adj MS	F-value	P-value
Regression	3	0.0.35	0.012	2.27	0.129
Orientation	1	0.031	0.030	5.96	0.030
Resolution	2	0.005	0.002	0.45	0.647
Error	13	0.067	0.005		
Total	16	0.101			

3.4 Summary

In this chapter, case studies have been used to illustrate the multi-objective optimization (MOO) approach as a strategy for quality achievement of the fused deposition modelling process. Time of printing, surface and mechanical integrity are the main objectives for the MOO process. Some of the MOO techniques which have been used in the FDM process are NSGA-II, Taguchi-Grey relational degree, Particle swarm optimization (PSO) and Pareto optimization methods. Usually, the decision variables in the FDM process are the 3D printing parameters such as temperature, speed, infill density, build orientations, layer thickness, among others.

References

1. N. Gunantara, A review of multi-objective optimization: Methods and its applications. Cogent Eng. **5**(1), 1–16 (2018)
2. L.S. de Oliveira, S.F.P. Saramago, Multiobjective optimization techniques applied to engineering problems. J. Braz. Soc. Mech. Sci. Eng. **32**(1), 94–105 (2010)
3. C.A.C. Coello, Multi-objective optimization. In: *Handbook of Heuristics* (Springer International Publishing, Cham, 2018), pp. 1–28
4. O.L. de Weck, Multiobjective optimization : History and promise. In: *Proc. 3rd China-Japan-Korea Joint Symp. Optimization Structural Mech. Syst. Invited Keynote Paper GL2-2* (2004), p. 14
5. C. Saule, R. Giegerich, Pareto optimization in algebraic dynamic programming. Algor. Mol. Biol. **10**(1), 1–20 (2015)
6. E. Asadollahi-Yazdi, J. Gardan, P. Lafon, Multi-objective optimization of additive manufacturing process. IFAC- Pap. Online **51**(11), 152–157 (2018)
7. Z. Jiang, Y. Liu, H. Chen, Multi-objective optimization of process parameters for biological 3D printing composite forming based on SNR and grey correlation degree. 3–8 (2015)
8. M.A. Matos, A.M.A.C. Rocha, A.I. Pereira, Improving additive manufacturing performance by build orientation optimization. Int. J. Adv. Manuf. Technol. (2020)
9. A. Dey, D. Hoffman, N. Yodo, Optimizing multiple process parameters in fused deposition modeling with particle swarm optimization. Int. J. Interact. Des. Manuf. No. 0123456789 (2019)
10. R. Ceylan, H. Koyuncu, ScPSO-based multithresholding modalities for suspicious region detection on mammograms. In *Soft Computing Based Medical Image Analysis* (Elsevier, 2018), pp. 109–135

11. S. Lalwani, S. Singhal, R. Kumar, N. Gupta, a comprehensive survey: applications of multi-objective particle swarm optimization (Mopso) algorithm. Trans. Comb. ISSN **2**(1), 2251–8657 (2013)
12. C.A. Coello Coello, G.T. Pulido,M.S. Lechuga, Handling multiple objectives with particle swarm optimization. IEEE Trans. Evol. Comput. **8**(3), 256–279 (2004)
13. N. Kim, I. Bhalerao, D. Han, C. Yang, and H. Lee, Improving surface roughness of additively manufactured parts using a photopolymerization model and multi-objective particle swarm optimization. Appl. Sci. **9**(1) (2019)

Chapter 4
Surface Engineering Strategy

Abstract In this chapter, surface engineering techniques for enhancing the surface properties of 3D printed parts are described. The strategies are classified broadly into traditional and modern (advanced) technologies. Under traditional methods, sanding, polishing, painting, gap filling and dipping are discussed as some of the processes for enhancing the surface quality of FDM parts. Under modern technologies, chemical vapour deposition, physical vapour deposition and vapour smoothing are identified as some of the most important methods for surface engineering of FDM parts. For each method, operating principles, advantages and limitations are described while quality enhancement aspects for FDM parts are also highlighted.

Keywords Chemical vapour deposition (CVD) · Vapour smoothing · Painting · Physical vapour deposition (PVD) · Polishing · Sanding · Surface engineering

4.1 Introduction

The major challenge with fused deposition modelling is the surface roughness and microstructural defects. The process is also limited by several printing errors which lead to poor surface quality. Most of these errors are learnt through experience. Some of these errors are caused by the misalignments of the nozzle and the build plate. The build plate only moves in Z-direction in fused deposition modelling and the step movements of the plate determine the layer thickness of the FDM parts. When the platform is not aligned properly, the prints cannot adhere properly to the surface since there occurs what is commonly known as the 'Elephant's foot' or the warpage syndrome. When the platform is misaligned, the finest details of the samples are usually difficult to capture in the printed product. Since the alignment of the platform determines the layer thickness accuracy, any misalignment significantly affects the roughness and patterning of the surface structure. To correct misalignment errors, the threaded screws of the printer should be checked for bending, and they should be wiped off any dust and greased frequently to reduce tear and wear.

The misalignment of the nozzle, on the other hand, is caused by loose belts, grub screws or debris blocking its smooth movement along the surface of the build plate.

Fig. 4.1 Illustrating shifted
layers during fused
deposition modelling of PLA
samples. The black arrow
shows the misalignment
between two consecutive
layers caused by shifted
layers

In such cases, the printer misses the bed and as such the print layers do not arrange very well. As such, the printed product will have some missing layers. The nozzle misalignment may also cause shifting of the layers as shown in Fig. 4.1. The shifting may lead to inaccurate form and size of the parts and such parts may not perform their functions as expected. For instance, if the 3D printed part is supposed to fit with another component, the shifted edges will not fit as expected. For a component printed as part of an investment, casting mould would not be accepted since such prints require a high level of precision in terms of form and dimension on their surfaces. Parts with shifted edges or layers cannot be accepted for biomedical, microelectronics and other advanced application sections.

Another significant error in FDM is observed at the starting point of the printing process. Usually for non-heated bed, the very first extruded material which comes into contact with the build plate has a tendency to cool earlier before touching the surface and it causes adhesion problems. Our experience has shown that it is important to isolate or cut off the first extruded material to avoid causing damage to the printed sample. Figure 4.2 shows an example of the incomplete fusion at the beginning of a print on our WANHAO desktop duplicator D10.

In our recent paper, titled 'Visual assessment of 3D printed elements : A practical quality assessment for home-made FDM products', presented in an international conference and published in Materials Today Proceedings, it was illustrated that the dimensional discrepancies between the CAD model and the 3D print of the most common geometrical attributes are due to the surface defects such as cracks, porosity, lack of enough fusion and adhesion [1]. These defects are usually manifested through poor surface quality and high roughness values. The defects occur within critical sections of the FDM print and such parts include corners, twisted sections, holes, sharp edges, very thin sections and so many others. An example of such defects on the rectangular feature from Fig. 4.2 is illustrated in Fig. 4.3.

Another major challenge during FDM of different materials, especially the flexible and graphene-based PLA filaments, is the loss of adhesion on the build plate. Whenever large prints are being manufactured and there is poor adhesion of the product on

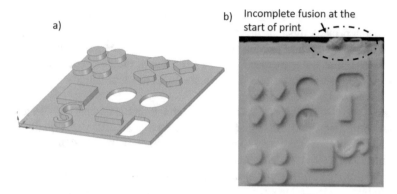

Fig. 4.2 **a** The CAD design for the features and **b** illustrating the incomplete melting and fusion at the beginning of the printing process (adapted from Mwema et al. [1])

Fig. 4.3 Defects due to insufficient fusion and melting on the corners and surfaces of the rectangular features. The defects cause poor quality surfaces on the FDM products

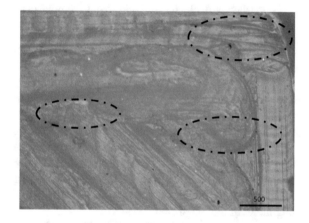

the plate, the back and forth action of the printer nozzle is likely to crash the entire printed sample. The problem is caused by titled platform, insufficient extruder and nozzle temperatures, uncalibrated machine, dirty build plate, and interference of the first unmolten extruded material with the raft or brim. The lack of adhesion may lead to warping and hence, failure of the printing process. In such cases, the layer alignment becomes inconsistent leading to poor surface quality and strength of the printed product. The problem can be corrected by undertaking the following:

- Undertaking bed levelling: This is usually a simple procedure, and for most desktop FDM printers, the manufacturers provide a methodology for undertaking the bed levelling. The idea is to set the minimum nozzle-bed plate distance, which is usually about 0.1 mm for most desktop 3D printers.
- Using adhesives or adding texture to the build plate: These are the two common ways most experts in the additive manufacturing community use to solve the problem of lack of adhesion on non-heated beds. Materials such as PLA do not

have major adhesion problems although when very fine details and large samples are being printed, this may arise. For example, in our WANHAO D10 Duplicator desktop printer, when printing with PLA, we apply general-purpose adhesives and Bostik Inc. adhesives have proven to be one of the most effective brands for us. Additionally, most of the 3D printers' beds are fabricated with honeycomb texture to enhance adhesion.

- Using heated build plates: Most flexible filament materials such as ABS and TPE must be printed on heated platforms for effective adhesion and quality of the prints. The temperature may vary based on the material type and model of the 3D printer. The user must determine the best table temperature for the material they are printing to ensure accurate adherence to their prints. For example, the typical plate temperature ranges for PLA, ABS, and PETG are 20°–60°, 80°–110° and 50°–75°, respectively. Usually, when the use of adhesives does not work, it is recommended to opt for a printer which has heated bed capability.
- Cleaning of the build plate: It is always recommended that the printing table should be washed with clean water and soap to remove previous adhesives or print debris to ensure sufficient adhesion. Additionally, an acetone moist cloth should be used to wipe off any dust. The table should also be dried off of any moisture before the printing process can be commenced.

The untimely depletion of filament wire during printing is another contributor to surface defects in the FDM process. It is advisable to ensure that there is enough spool of filament wire before starting the printing of parts. External shocks causing vibrations to the printer are also a contributor to defective surfaces in 3D printing. Usually, a 3D printer (desktop) should be placed on a stable table away from unnecessary movements and human activities which may interrupt the action of the printer. Vibrations may cause shifting of the workpiece during printing, which results in misalignment of the layers and may also be detrimental to the nozzle and the entire alignment of the 3D printer.

The quality of the printed products largely depends on the printer settings as it has been described in the previous chapters. As described, these parameters directly influence the extrusion, layer fusion and adhesion of the print onto the build plate [2]. The proper choice of these parameters is paramount for the quality of the surfaces. Optimization strategies have been described earlier as strategies for enhancement of the quality of 3D printed products. In this chapter, we represent surface engineering as a strategy for quality enhancement of fused deposition modelling components. Here, surface engineering may refer to finishing processes as well as functionalization of the surfaces of the 3D printed parts [3]. We discuss both traditional and emerging technologies of surface engineering for FDM products with emphasis on best practices for adoption to quality manufacturing.

4.2 Traditional Surface Engineering Methods for FDM Products

4.2.1 Sanding and Polishing

This is usually the first finishing operation undertaken on the FDM products after the supporting material (brim or raft) has been removed. The process involves grinding the surface of FDM parts with an abrasive paper, usually Silicon carbide (SiC) papers are commonly used in metallographic processes. There are various grades of the grit (SiC abrasive materials) and the choice of the range through which the FDM parts are to be ground depends on the layer resolution and surface quality of the printed part. For surfaces, with very clear surface defects or blemishes, the sanding should start with the roughest SiC paper, which is usually 100 grit, followed by 220, 400, 600, 1000 and 2000 grit sandpaper in that order. Usually, the sandpaper should be kept wet to avoid excess heat generation and build-up of debris which further deteriorates the surface quality. It is highly recommended to clean the surface of the FDM part with a soft brush (such as toothbrush) and soapy water and dried with a smooth cloth to prevent any dust build-up and 'caking'. To achieve very shiny surfaces, the sanding can be extended beyond 2000 to 5000 SiC grade.

There are several ways in which sanding can effectively be undertaken on the FDM parts. Sandblasting is one of the interesting techniques to achieve consistent surfaces on FDM parts. The process is undertaken in the normal sandblasting equipment (shown in Fig. 4.4) in which the blasting media (sand or any other grit material) of different grades can be used [4]. The process involves accelerating the grit material (sand) through a nozzle to the surface of FDM printed parts. The best practice is that after the sandblasting process, the samples are washed in soapy water, dried in high-pressure air and wiped off with a dry paper towel. Drying with ambient air is strongly discouraged since it causes spotting on the surfaces of PLA prints after sandblasting. The main disadvantage of sandblasting is that it changes the surface colour of most printed materials, for example, as shown in Fig. 4.5, white PLA sample changes to a bone hue colour whereas the black PLA changes to a nice matte grey.

Sanding offers the following advantages:

i. It is a simple process which can be done through handheld SiC papers as shown in Fig. 4.6. It is therefore easy to monitor although that may depend on the experience of the user.
ii. It can also be done on belt sanders or on improvised rotating machines such as workshop lathe, wooden machines or even rotating parts of a bicycle.
iii. The method is cheap and easy to carry out on the prints.
iv. It can be undertaken nearly on all parts, except very tiny or thin parts. When very large parts are to be smoothened, sandblasting should be used.
v. Sanding is effective in removing the staircasing effect of the FDM process on parts.

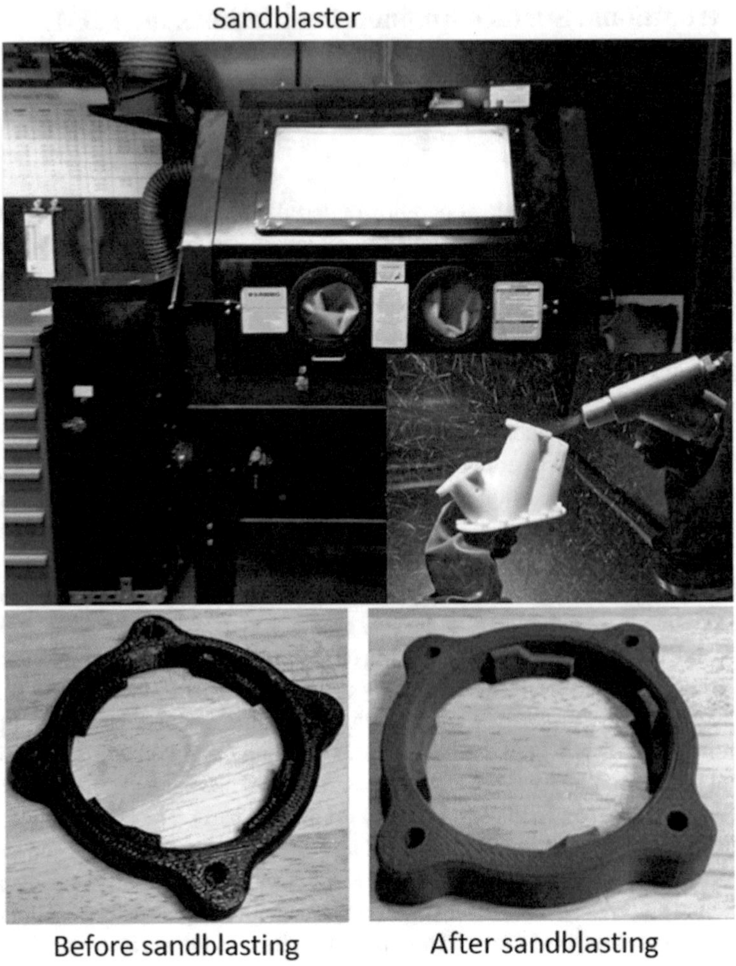

Sandblaster

Before sandblasting   After sandblasting

Fig. 4.4 Sandblasting of FDM parts. The surface quality transformation after sandblasting is also shown (obtained from open-access webpage: www.instructables.com). The inset illustrates the process of the sandblasting inside the blasting chamber

The method is however limited by the following aspects:

i. It is difficult to control the material removal rate, especially when doing hand sanding.

ii. It is not easy to do hand sanding on very thin-walled samples as they may collapse or even hurt the fingers of the individual undertaking the sanding. It is also very difficult to sandpaper complex shapes and features such as tiny holes and twisted structures.

Fig. 4.5 Colour transformation on prints prepared from black and white PLA filaments after sandblasting under similar conditions (obtained from open-access webpage: www.instructables. com)

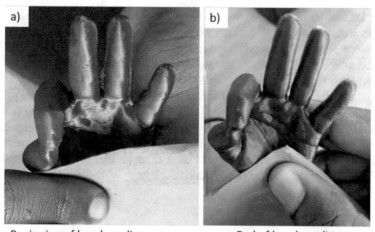

Fig. 4.6 Illustrating hand sanding of brown PLA OK sign using SiC grinding paper (#1200). Colour transformation can also be observed after the sanding process from **a** original brown PLA to **b** whitish surface after PLA

iii. The colour change during sandblasting may be undesirable for some applications, especially for decorative purposes. It may call for unnecessary painting to restore the desired filament appearance.

Vibratory grinding is a related technology for finishing FDM parts in mass processing. In this process, the parts are ground by vibrating the parts inside a bowl or tub containing the suitable grit media. Some of the most common media available from

Fig. 4.7 Examples of grinding vibratory machines used in the finishing of FDM parts. Obtained from Stratasys Ltd. webpage (www.stratasys.com)

the market include ceramics, synthetic, plastic and maize cobs specially prepared by different trademarks such as Stratasys® limited. Figure 4.7 shows pictures of some of the common grinding vibratory machines for FDM finishing.

After sanding and blasting, the FDM parts can be exposed to the polishing process to achieve the long-lasting, fine and mirror-like finish. The process involves buffing the surface with a buffing wheel or by using a cloth with embedded particles (microscale) together with a polishing chemical. As known, Blue Rogue is one of the chemicals extensively used for the polishing of pieces of jewellery made from plastics and synthetic materials. The process is similar to that used in cleaning car headlights and related chemicals can be used although care should be exercised to avoid a chemical attack of the print material.

4.2.2 Painting and Priming

Painting and priming are finishing processes which are undertaken mostly after sanding of the FDM parts' surfaces. They are undertaken to further enhance the surface quality and appearance of the prints. Priming is undertaken first before the painting and it is meant to remove any noticeable surface flaws. The most important aspect of priming is to choose a primer which is compatible with the FDM material and using a sprayer to apply the primer to ensure uniform coating. For the effective application of the primer, the following points are important to take into consideration:

i.   First, ensure the primer is well mixed before spraying on the printed part. The mixing procedure will always be provided under the manufacturer's instructions.

ii. Using an aerosol sprayer, two thin layers should always be sufficient for the priming of FDM parts. While doing so ensure uniform coverage of all the details and that the primer does not obscure some of the finer details of the parts.
iii. The spraying should be done while approximately at a distance of 20 cm away from the part to ensure the pooling of the primer. It is important to continuously rotate the part while undertaking this process for uniform distribution of the primer layers.
iv. After the first layer of the primer, always inspect the part to see if there is any need for sanding or rebuilding some features in case there is a loss of some details. If any sanding is necessary, wait for the coat to dry and then sand the necessary sections of the part using 600 grit SiC paper.
v. The last coating of the primer should be very fine and this is achieved by sweeping across the part with rapid bursts of the spray. This method ensures that there is no building up and dripping of the primer.

Once the primer is dry, the painting can commence. Usually, acrylic paints and brushes used by the artists are effective, although the use of airbrush or aerosol can offer better surfaces. The process gives better results if very thin layers are applied consecutively until it forms an opaque layer. Some important points to enhance the quality of the painting are listed below.

i. A painter's tape is very important at this stage to mask the areas of the print which should not be painted or which may require a different colour paint.
ii. When painting, rotate the parts quickly to ensure uniform coating of the paint on the surface of the parts. The first layer of paint should be quickly done and left for 10 min to avoid sagging.
iii. The finish coat should be applied as the last layer. This coat is applied to protect or seal the paint surface and to maintain the desired sheen. It is advisable to polish this layer with wax or any other polishing chemical as directed by the manufacturer of the paint. The paint should be left to dry for at least 12 h after which the surfaces are cleaned with a tack cloth.
iv. The processes should be undertaken in a well-controlled environment in which there is minimal influence by the environment such as air. Wind can easily deposit particulate matter on the surfaces of the parts during painting. These particles get trapped under the paint and may cause later delamination of the paint.
v. The best practice is to undertake the painting in one session for the effective distribution of the paint on the surface. Multiple coats are preferred than a single coat but each coat should be as thin as possible.

The common important aspects to consider during these processes to achieve the quality surfaces are the choice of the type of primer, paint and tools (including brushes), and the curing times. During painting, it is advisable to undertake polishing so that shiny and uniform surfaces can be achieved.

4.2.3   Gap Filling

During 3D printing, gaps, voids or cracks may occur on the surfaces due to nozzle constraints and may lead to defective products. For instance, for FDM parts to be used as moulds for investment casting or injection moulding of microelectronics applications, such defects are undesirable. Gap filling involves using an epoxy or other materials such as automotive body fillers to cover these voids or cracks. Once the defects have been filled, sanding or polishing may also be necessary to achieve the desired form and dimensional accuracy. One of the most advanced procedures for gap filling is vacuum infiltration [5]. In this process, the parts are submerged in a sealant and exposed to the vacuum chamber, usually about 0.375 torr. Some of the most common sealants for vacuum infiltration include oil-based polyurethane, phenolic resin, acrylic resin, epoxy resin and so forth [5, 6]. The method is effective in reducing surface defects since the sealants flow into the internal gaps and voids; the sealants harden within these defects (voids and gaps) thereby filling and reduce them on the 3D printed part. The process enhances surface smoothness, adhesion between raster and does not mostly affect the dimensions of the FDM part. The effect can be related to carburization of metals [7] and also enhances the mechanical strength and water absorption characteristics of the FDM parts [8].

4.2.4   Dipping

The process involves the coating of the FDM parts by dipping them into the coating medium. The dipping is meant to enhance the surface quality (reduce surface roughness) or/and functionalize the surface of the printed parts. The coating media is filled into a container based on the size of the printed part such that the part can be fully covered by the media on dipping without splashing which may lead to spooling. During this process, the most important aspect is the time of dipping as it determines the thickness and absorption or reaction of the coating and printed materials. Other parameters affecting the dipping process include coating viscosity and the number of dips of the print into the coating media. This technique has been utilized by various researchers to improve the surface quality and performance of FDM printed parts.

A study by Lee et al. [9] is superhydrophobic 3D printed PLA through dipping in hydrophobic fumed silica nanoparticles. The dipping process was undertaken for 1 min and the samples dried at room temperature for 12 h and then cleaned in an ultrasonic bath containing ethanol as the media for about 10 min. The samples were then dried for 1 h at room temperature and evaluated for hydrophobicity and mechanical strength. The dip-coating process was shown to enhance the superhydrophobic characteristics of the complex 3D printed parts although there was a slight decrease in tensile strength, tensile modulus and elasticity of the parts. In another study, the influence of coating speed, drying conditions and the number of coated layers in computer-controlled dip-coating process of PLA were investigated and their effects on surface roughness of the printed parts reported [10]. Water-based polyurethane coatings were used in this study from two commercially available brands (Pro Finisher, a clear gloss polyurethane, mostly used for wood floor finishing and JetFlex® polyurethane dispersion mostly used for plastic coatings in aircraft interior). The study revealed that these coatings were able to fully cover the curved surfaces of the printed parts and lower their surface roughness.

In our recent project (unpublished), we have dip-coated PLA printed samples with conductive films of silver for enhanced surface properties and applications of the samples. Upon dip coating, the samples were then heat-treated at a constant temperature of 100 °C and at different exposure times of 0, 5, 10 and 20 min. We undertook several characterizations on surface topography, morphology and structure, and the following important conclusions were drawn:

i. The surface roughness: Through the silver coating, it was possible to lower the surface roughness to nanoscale levels and the roughness evolved with the heat treatment of the PLA substrates. The root mean square roughness values for untreated samples were the lowest (around 117 nm) whereas the highest roughness was obtained on the surface of the samples treated for 20 min (174 nm). Usually, the surface roughness of PLA printed surfaces is obtained through optical profilometry since it is usually on a macroscopic scale. In this case, the silver dip-coating surfaces were able to be measured using atomic force microscopy (AFM) and the images are shown in Fig. 4.8. These results are illustrative of the role of metallization and surface treatment on the surface roughness of the PLA printed parts.

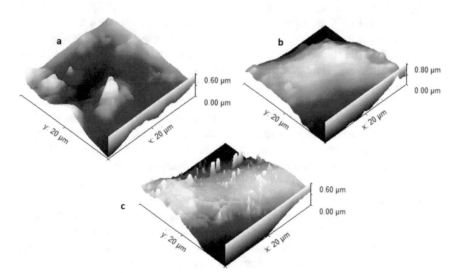

Fig. 4.8 Three dimensional (3D) atomic force microscope images of PLA-silver coated samples indicating reduced surface roughness. The samples shown are heat-treated for **a** 0 min **b** 10 min and **c** 20 min at constant temperature of 100 °C

Fig. 4.9 The low-resolution
SEM on the surface of
untreated PLA-Ag samples.
The staircasing effect of the
FDM process was
considerably reduced

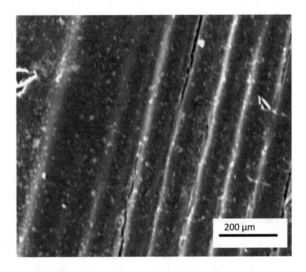

200 µm

ii.   Surface morphology: We undertook both low-resolution and high-resolution
      scanning electron microscopy on the PLA-Ag heat-treated samples and it was
      interesting to observe that the surface cracks and porosity evolved with the
      heat treatment. The low-resolution SEM revealed a very low presence of the
      staircasing effect usually observed on 3D printed parts as shown in Fig. 4.9. The
      high-resolution SEM showed that on heat treatment of the PLA-Ag samples,
      there occurred infusion between the substrate and dip coating of silver.
iii. XRD and XPS analyses: A further analysis on the surface of Ag-PLA samples
      exposed to 100 °C at different times revealed important chemical and structural
      transformations necessary for influencing the surface quality of the printed parts.
      Crystalline structures of silver coating were observed with a face-cantered cubic
      (FCC) structure. The intensity of the XRD peaks was seen to evolve with heat
      treatment. The heat treatment also led to the transformation of PLA structure
      from δ (unstable) into α (stable) structure. The recrystallization of the PLA on
      exposure to 100 °C is the reason for the reduction in porosity and infusion of the
      silver coating into the PLA structure. The XPS was undertaken to investigate
      the binding energy of the silver metal onto the PLA surface. The XPS results
      confirmed bonding between PLA and silver on heat treatment.

Dipping can be explored as an alternative for vapour deposition techniques in cases
where the thickness of the films is not significant; dipping cannot achieve the level
of nanoscale obtained through chemical and physical vapour deposition methods.
As mentioned, the quality of the coating depends on the time of dipping, chemistry
of the coating media, speed of dipping and so many other parameters. Therefore,
optimization studies are necessarily based on the available technology for the dip-
coating process.

4.3 Advanced Surface Engineering Methods for FDM Products

4.3.1 Chemical and Physical Vapour Depositions

These are the most advanced techniques for surface engineering and improvement of quality of surfaces in various fields. The processes deposit very thin layers of coatings (known as films) on the surface of the printed parts. The depositions are undertaken for two reasons:

i.   To reduce the surface roughness of the 3D printed samples by filling into the microporosity and microcracks on their surfaces.
ii. To enhance or impart additional functionality of the printed parts. For example, for the microelectronics industry, flexible FDM parts may be coated with conductive coatings to enhance their electrical conductivity. In biomedical applications, PLA printed samples can be coated with hydroxyapatite or titanium carbide (TiC) to enhance their surface strength as well as biocompatibility properties.

There are two major methods of depositing very thin layers of coatings on surfaces, classified as chemical vapour or physical vapour deposition methods. In chemical vapour depositions (CVD), chemical processes are involved to create the compounds for thin coatings whereas, in physical vapour depositions (PVD), the depositing material is physically removed from the source material and deposited to the surface to be coated. These processes have been schematically illustrated in Fig. 4.10. The choice of any of the process depends on the application and type of coatings required for the FDM parts.

In both methods, the surface modification enhances both the quality of the surface and performance. For example, polydopamine and type I collagen coatings on PLA printed samples for biomedical scaffolding were shown to enhance their mechanical properties as well as improve their surface roughness [11]. Such coatings also enhance the bioactivity of the PLA components.

The surface modifications through PVD and CVD also play an important role in altering the wettability properties of the surface of the PLA and other printed materials. For example, to enhance hydrophobic or superhydrophobic properties of printed parts, CVD can be used to deposit film materials such as silica particles and complex-based hydrocarbons onto the printed parts. In such cases, the coated 3D prints have a higher contact angle when water contact tests are undertaken on them as shown in Fig.4.11 [12].

One of the most attractive aspects of the physical vapour deposition methods is their capability to create thin films of various materials including a wide range of metals [13]. As such, techniques such as sputtering and thermal evaporation are very attractive in the metallization of PLA, ABS, rubber-based and so many other printed materials. For instance, Juarez et al. [14] utilized a magnetron sputtering method to deposit pure aluminium, Inconel 600 and Ti6Al4V on PLA printed micro-trusses.

Such studies have the potential to expand the applications of polymer micro-trusses functionalized with metal thin films. Metallization of FDM parts is generally meant to achieve the following:

i.   Enhance the surface strength of the FDM materials such as PLA, ABS and so forth [15].
ii. Since conductive filaments such as graphene-doped PLA are challenging to print on most readily affordable desktop printers, metallization imparts the conductivity capabilities of the printed parts.
iii. The surface deposition of metallic thin films is also meant to enhance specific chemical characteristics such as corrosion resistance to the printed part.
iv. The metallized thin films from PVD and CVD techniques are in the form of atomic layers and usually in terms of nanoscale. In fact, thin films have their roughness values in the ranges of nanoscale. As such, thin-film metallization lowers the surface roughness of the printed parts considerably. Therefore, lower roughness is achieved through these processes.

The fact that nanotechnology and the need for intricate micro-devices in the modern engineering communities are exponentially rising implies that continued synergy of vapour deposition techniques and FDM (or other AM techniques) is of no doubt. Therefore, there are growing research activities for the advancement of these strategies for quality enhancement of fused deposition technology in the following aspects:

i.   Both CVD and PVD depend on various processing parameters such as temperature, equipment pressure, flow rates of the chemicals and fluids, quality of the 3D printed surfaces, chemistry of the materials among many other parameters. For the strategy of surface engineering to work effectively for 3D printed parts, proper combinations of these parameters are necessary. As such, optimization strategies of PVD and CVD processes for depositions of thin films on 3D printed parts are very critical. Already such optimization studies on technologies such as sputtering for non-printed substrates are available [16, 17].
ii. Adhesion of thin metals/alloy compounds onto most FDM materials is a big challenge due to the lattice differences. There are various strategies to ensure that the thin films adhere and one of them is heat treatment of the printed parts after thin-film deposition. When this is done, the printed material flows locally to bind with the thin layer. Another way is to introduce an interlayer between the functional thin layer and the printed substrate. Etching of the substrate surface using chemicals can also enhance bonding between the metal and FDM material. Surface etching increases the surface area, creates more sites for anchorage by the metallic films and cleans the substrates off any surface contamination [18].
iii. To expand the applications of 3D printing in various industries, there is a need to evaluate the depositing behaviour of a wide range of materials for thin films. The capability of the FDM to produce intricate features is attractive for complex circuit and sensor systems and therefore, CVD and PVD techniques present a golden opportunity to harness the full potential of the FDM.

4.3.2 Vapour Smoothing

Figure 4.12 shows the typical vapour smoothing equipment and as shown it has two chambers namely the smoothing and cooling chambers and it has been proven successful in ABS materials [19]. The smoothing chamber contains the smoothing fluid which is usually of boiling temperature of 43 °C. An example is a fluid consisting of 30% of Decafluoropentane and 70% of Trans-Dichloroethylene [20, 21]. Acetone with an assay of 99% has also been used as a smoothing fluid for ABS printed parts [20, 22]. The cooling chamber is connected to a refrigeration system which keeps the temperature to as low as 0 °C. The vapourized fluid from the smoothing chamber gets to the cooling chamber and is condensed for recirculation such that there is no need to refill the smoothing fluid. The 3D printed parts are usually suspended above the smoothing chamber such that once the fluid is vapourized, it impinges on its surfaces. The smoothing fluid is heated to 65 °C and rises to the printed part as hot vapour [23]. While on the surface of the printed part, the hot air causes the material to flow such that the surface flaws and rough steps get filled, and hence smoothen. The time of exposure and other settings of the vapour smoothing equipment affect the smoothening characteristics of the printed part [20].

The vapour smoothing technique helps in eliminating the staircasing effect of the FDM parts. It has been reported in the literature that when acetone is used as a smoothing vapour for ABS parts, there is a creation of ABS-acetone slurry through which material equal to the staircase steps on the parts is removed [22]. The challenge in this process is determining the accurate exposure time under which the roughness is eliminated without further damage to the surface of the printed part. When correct times are determined, the obtained value of roughness is usually very low and comparable to the injection moulding process of the print material [24].

Although the vapour smoothing process improves the surface quality of the printed parts, it can cause defects on the surfaces of the prints. It has also been shown that

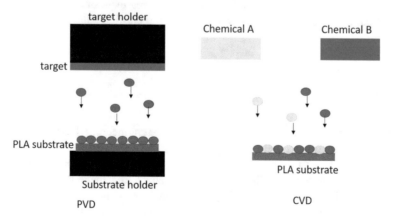

Fig. 4.10 Illustrating physical vapour deposition and chemical vapour deposition methods

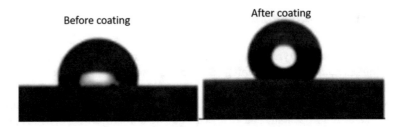

Fig. 4.11 Showing typical contact angle results of uncoated and coated 3D printed samples. Thin layer coating is used to enhance the hydrophobicity of 3D printed parts for different applications

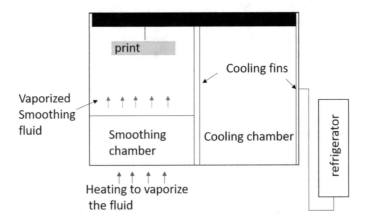

Fig. 4.12 Schematic diagram for vapour smoothing equipment [19]

it affects the thermal stability of the 3D printed ABS parts [25] and lowers the mechanical strengths of the ABS printed material [26] and it also increases the density of the printed materials [27]. The acetone treatment of the surface of ABS printed parts has been shown to lower hardness and elastic modulus of the material [28].

4.4 Summary

In this chapter, the modification of surfaces (surface engineering) has been presented as the strategy for quality enhancement of FDM parts. There are two broad categories for enhancing the surface of 3D printed parts, namely the traditional and advanced methods. The traditional methods are usually meant to achieve the better surface appearance and reduce the staircasing effect (roughness) on the FDM parts. On the other hand, advanced methods are mostly meant to functionalize the surfaces besides improving the surface quality of the 3D printed parts. The chapter has also highlighted some of the important aspects of ensuring that better surfaces are achieved for these methods based on existing literature.

References

1. F.M. Mwema, E.T. Akinlabi, O.S. Fatoba, Visual assessment of 3D printed elements : A practical quality assessment for home-made FDM products. Mater. Today Proc. (2020)
2. M. Pérez, G. Medina-Sánchez, A. García-Collado, M. Gupta, D. Carou, Surface quality enhancement of fused deposition modeling (FDM) printed samples based on the selection of critical printing parameters. Materials (Basel) **11**(8), 1382 (2018)
3. J.S. Chohan, R. Singh, K.S. Boparai, Effect of process parameters of fused deposition modeling and vapour smoothing on surface properties of abs replicas for biomedical applications. In *Additive Manufacturing of Emerging Materials* (Springer International Publishing, Cham, 2019), pp. 227–249
4. I. Gajdoš, E. Spišák, L. Kaščák, V. Krasinskyi, Surface finish techniques for FDM parts. Mater. Sci. Forum **818**(May), 45–48 (2015)
5. J. Mireles et al., Analysis of sealing methods for FDM-fabricated parts. In: 22nd Annual International Solid Freeform Fabrication Symposium—An Additive Manufacturing Conference, SFF 2011 (2011), pp. 185–196
6. J.G. Zhou, M. Kokkengada, Z. He, Y.S. Kim, A.A. Tseng, Low temperature polymer infiltration for rapid tooling. Mater. Des. **25**(2), 145–154 (2004)
7. E.Y. Salawu, O.O. Ajayi, A.O. Inegbenebor, S. AkinlabI, E. Akinlabi, A.P.I. Popoola, U.O. Uyo, Investigation of the effects of selected bio-based carburising agents on mechanical and microstructural characteristics of gray cast iron. Heliyon **6**(2) (2020)
8. M. Miguel, M. Leite, A.M.R. Ribeiro, A.M. Deus, L. Reis, M.F. Vaz, Failure of polymer coated nylon parts produced by additive manufacturing. Eng. Fail. Anal. **101**(January), 485–492 (2019)
9. K.M. Lee, H. Park, J. Kim, D.M. Chun, Fabrication of a superhydrophobic surface using a fused deposition modeling (FDM) 3D printer with poly lactic acid (PLA) filament and dip coating with silica nanoparticles. Appl. Surf. Sci. **467–468**, 979–991 (2019)
10. J. Zhu, J.L. Chen, R.K. Lade, W.J. Suszynski, L.F. Francis, Water-based coatings for 3D printed parts. J. Coatings Technol. Res. **12**(5), 889–897 (2015)
11. E. Baran, H. Erbil, Surface Modification of 3D printed PLA objects by fused deposition modeling: a review. Colloids and Interfaces **3**(2), 43 (2019)
12. C. Cheng, M. Gupta, Surface functionalization of 3D-printed plastics via initiated chemical vapor deposition. Beilstein J. Nanotechnol. **8**(1), 1629–1636 (2017)
13. F.M. Mwema, O.P. Oladijo, S.A. Akinlabi, E.T. Akinlabi, Properties of physically deposited thin aluminium film coatings: A review. J. Alloy. Compd. **747**, 306–323 (2018)
14. T. Juarez, A. Schroer, R. Schwaiger, A.M. Hodge, Evaluating sputter deposited metal coatings on 3D printed polymer micro-truss structures. Mater. Des. **140**, 442–450 (2018)
15. A. Equbal, A. Sood, Metallization on FDM parts using the chemical deposition technique. Coatings **4**(3), 574–586 (2014)
16. O. Abegunde Olayinka, A. Esther, O.P. Oladijo, J.D. Majumdar, Surface integrity of TiC thin film produced by RF magnetron sputtering. Procedia Manuf. **35**, 950–955 (2019)
17. O.O. Abegunde, E.T. Akinlabi, P.O. Oladijo, Dataset on microstructural, structural and tribology characterization of TiC thin film on CpTi substrate grown by RF magnetron sputtering. Data Br. **29**, 105205 (2020)
18. A. Equbal, A.K. Sood, Investigations on metallization in FDM build ABS part using electroless deposition method. J. Manuf. Process. **19**, 22–31 (2015)
19. C.-C. Kuo, C.-M. Chen, S.-X. Chang, Polishing mechanism for ABS parts fabricated by additive manufacturing. Int. J. Adv. Manuf. Technol. **91**(5–8), 1473–1479 (2017)

20. R. Singh, S. Singh, I.P. Singh, F. Fabbrocino, F. Fraternali, Investigation for surface finish improvement of FDM parts by vapor smoothing process. Compos. Part B Eng. **111**, 228–234 (2017)
21. D. Singh, R. Singh, K.S. Boparai, Investigations on hardness of investment-casted implants fabricated after vapour smoothing of FDM replicas. J. Brazilian Soc. Mech. Sci. Eng. **42**(4) (2020)
22. A. Lalehpour, C. Janeteas, A. Barari, Surface roughness of FDM parts after post-processing with acetone vapor bath smoothing process. Int. J. Adv. Manuf. Technol. **95**(1–4), 1505–1520 (2018)
23. J.S. Chohan, R. Singh, Enhancing dimensional accuracy of FDM based biomedical implant replicas by statistically controlled vapor smoothing process. Prog. Addit. Manuf. **1**(1–2), 105–113 (2016)
24. A. Colpani, A. Fiorentino, E. Ceretti, Characterization of chemical surface finishing with cold acetone vapours on ABS parts fabricated by FDM. Prod. Eng. **13**(3–4), 437–447 (2019)
25. S.-U. Zhang, J. Han, H.-W. Kang, Temperature-dependent mechanical properties of ABS parts fabricated by fused deposition modeling and vapor smoothing. Int. J. Precis. Eng. Manuf. **18**(5), 763–769 (2017)
26. A. Garg, A. Bhattacharya, A. Batish, Chemical vapor treatment of ABS parts built by FDM: Analysis of surface finish and mechanical strength. Int. J. Adv. Manuf. Technol. **89**(5–8), 2175–2191 (2017)
27. J.S. Chohan, R. Singh, K.S. Boparai, Post-processing of ABS replicas with vapour smoothing for investment casting applications. Proc. Natl. Acad. Sci. India Sect. A Phys. Sci. (2020)
28. V.T.L. Costa, C.N. Pai, Superficial characteristics of acetone vapor treated ABS printed parts for use in upper limb prosthesis. IFMBE Proc. **70**(1), 365–376 (2019)

Printed in the United States
By Bookmasters